Astrology made easy

Pisces Venus 27° Meena JUPITER	Aries Sun 10° Mesha MARS	Taurus Moon 3° Vrishbha VENUS	Gemini Mithun MERCURY
Aquarious Kumbh SATURN			Cancer Jupiter 5° Kataka Moon
Capricorn Mars 28° Makar SATURN			Leo Simha SUN
Saggittariu Dhanush JUPITER	Scorpio Rahu/Ketu Vrishthika MARS	Libra Saturn 20° Tula VENUS	Virgo Mercury 15° KANYA MERCURY

Exaltation - Planets in Red
Name of Rashi in English & Hindi in Black
House owned by Planet in Blue.

2015/12/18 15:36

Rashi Chart

Jaipal Singh Datta

DEDICATION

This book is dedicated to my Guru H.R.Seshadri Iyer of Bangalore, India. I never believed in Astrology. He convinced me and proved that Astrology is a science. He predicted that I shall spread his knowledge around the world and spread his name. In this book i.e Astrology Made Easy, I have tried to share information and knowledge received from my Guru. This is Astrology Part (1) book as per his nomenclature.

CONTENTS

ACKNOWLEDGMENTS

I am thankful to Brigadier (Retd) Dr. Kapil Mohan Padam Shree V.S.M. PhD and Smt. Pushpa Mohan for their encouragement for this novel subject. I am thankful to my parents, elders, Mohyals, and my global family. I am thankful to my children for their love and affection. I was lucky to have good parents, who took care of me. I am lucky to have good family, who stood by me in all odds. I am lucky to have good employer, who helped me and supported me for this noble cause. Now I understand that I am nothing. I am only puppet in the hands of destiny. May God help my readers to understand this subject on scientific basis? I am thankful to all my friends who loved me and love me.

I came to know Mr.H.R.Sheshadri Iyer while I was working as Chief Executive Officer in Artos Breweries Ltd, Ramachandrapuram, Dist. East Godavari, Andhra Pradesh. Mr. V.S.S.Sharma, excise inspector, introduced me to the writings of Mr. Sheshadiri Iyer. I met him in Bangalore in India. He lived in Kaveribai Layout, Appurao Road, Chamrajpet, Bangalore 560018. His phone number was Bangalore 601509 in 1980. He was coaching Astrology by direct and postal methods. He used to provide consultation to individuals and used to predict in writing. He has published many articles on Astrology, importance of division charts in Astrology, importance of Thithi or dates as per moon movement in Astrology Magazine published by Dr. B.V.Raman Bangalore. He tried to present Astrology in a very simple language, so that any layman can follow and understand. He is my Guru. My all respects to him.

With the passage of time, mathematical calculations of Astronomy became very easy due to computer technology. So it is time to understand Astrology as science. Due to this reason I am spending some money and time to spread this knowledge by Internet. Google, Microsoft and information technologists helped me to present my writings to others.

Thanks to my Guru and all my beloved teachers.

CHAPTER I   FUNDAMENTALS

We live in a society. Society is governed by social and national laws. Nation is governed by international law of the World .World is a part of universe and universe is part of Brahmand / Cosmos. Believe it or not Brahmand is also governed by certain laws. This is known as law of nature. Nature is ruthless. In India we believe in the law of nature and we always respect and worship nature.

Nature has five elements or known as TATWAS like air, water, earth, fire and energy / Akash. These all are governed by laws of Brahma. Study Geeta for Laws of Brahma. We can also name these elements as carbon, hydrogen, nitrogen, oxygen and factor which joins all these elements. It may be energy. Amino acids are part of these elements and amino acids are part of DNA or life. We know matter can neither be constructed nor destroyed. Total of energy is a constant or there is a perfect law for all this.

One can create matter from energy and energy from matter by brain power. Brain is a processing unit of computer in human beings. As computer is controlled by a program, in the same way human being is controlled by a program. That program is fixed by gravitational forces of Sun, Moon, Mars, Mercury, Jupiter, Venus, Saturn, Rahu and Ketu at the time of birth by rays of these planets and 27 stars. Angle of rays of these 9 planets varies every second. It also changes at different places as earth is rotating and moving. Due to this DESTINY of each individual is different. To know details of angles of the said planets

1

please read Ephemeris by Lahiri. NASA has helped in computing star positions or planet positions in the World and helped Astrologers on this account.

To study all this we need public help and patronage of the government. Bearing this in mind after a stage of study I took to research work and after a long time when my planets also were favorably disposed for genuine research some new flashes struck me and some advantageous situations arose of their own accord. When I say this I do this with authority. Even a well-read and proficient astrologer, if not backed by favorable planets, is likely to go down at such unpropitious times. This being Vedanga. All cannot hope to be successful at all times. One must be Godly and before venture he must think of his deity and then proceed. For, it is only then he will have proper flashes especially in matters of alternative situations.

Next I wish to trace the origin of this Science, its gradual development and its present state. All know that it is Vedanga. As Veda is spelled by God so is this Science too. There are versions stating that Iswara narrated this Science to Parvathi and Nandikeswara to his disciples then came the days of Maharishis who by their superior knowledge and clairvoyance taught their Disciples orally. You may be aware of one Saptha-Rishi Vakyam meaning that the seven Maharishis had discussed this science. Next came the days when some of their disciples took to scribing mostly on Palm-Leaves. So far this Sacred Science was being held safe in their hands. Next followed the days of Daivaganas who though not equal to the class of Rishi-Shisyas were of sufficiently superior intellect and more than all godly. Thus with their intellect and intuition they wrote out texts. Unfortunately it was Poetic Era, Prose being of recent origin. In the anxiety to maintain Rhyme and Grammar they had to use sometimes words with double meaning or even distracted Tien-. The author is sure of the proper meaning of his writings; but when others read his verses very likely doubts arise. Actually it has been so and this has given rise to many commentators viewing the same matter differently. Varahamihira in his Brihat-Jataka has in the very opening verse said that due to hurried writing and short space he had to compose verses with words of deep and multiple meanings-Any one perusing Brihat-Jataka may feel that there is nothing substantial helping prediction. One may even go to the extent of saying that his sayings may not be always true. I too held such a view so far. Yet this is the first text quoted by all.

Really it must be a great and valuable work. It is only when my research proved good I began to appreciate the monumental work. Very many secrets are there hidden. For example, a long chapter deals with various kinds of **Ayurdayas.** It is not meant only to estimate the longevity. From the several Rays attributed to planets the magnitude of effects may be gauged. I have illustrated them in my Chapter VII on finding Quantum of wealth and number of issues. Likewise I have seen Astrologers gauging other matters but they have held them with themselves and for themselves. If they come out with those secrets they would be not only rendering their meritorious service to this science but would be helping the public as well.

Astrology is a very useful subject which deserves to be introduced in universities. I am aware of the course of Astrology in Sanskrit Colleges. After a number of years of study one gets out as Jyothir Visharada but of what avail, he will not be able to handle a horoscope. The nature of this course is such that it is filled up mostly with theories than practice. What is needed is more of practice with applications of principles by which modern conditions may be estimated and this can be successful only at university level- Some may say that it is not a definite science to be introduced in colleges. If so, are other subjects like economics, Statistics, Philosophy, Psychology, Logic etc. perfect subjects? After all does it require one to waste his teens and part of youth to study History, Economics, Philosophy, Sociology, Politics, and Literature etc. at the college? In fact any one with the knowledge of letters will be able to understand them even in a much shorter time and better level than a college student reading for the sake of a Degree. Such a Degree holder may not be in a position to impart even that little knowledge to others once he gets out of the college as his immediate anxiety would be job-hunting. My Guru says "I say this with personal experience. What have I been able to do with my Mathematics M. A- Degree (with first class Graduate ship and Merit scholarship) — only to become a Government Servant for which an ordinary S. S. L- C will do as is now actually. It is this notion that made me study a Noble subject which may not only help me but be helpful to the public also. At this stage I feel to forget my M.A. Degree or say any degree of science or technology and be proud to be called a Post-Graduate in Astrology." After this little dilation let me go back to the subject.

Having surveyed the development of the Science up to the period of Daivaganas we have to know its condition later. Again two types of

3

Astrologers exist now- One of scholarly type who have held back the essence of this science with themselves and held it very secretive making business out of it and finally burying their knowledge as invariably their sons prove unworthy of their father to inherit this knowledge. The other type is what we find in plenty these days - one dollar Astrologers posing as Professors after reading a book or two on the subject. These latter class of astrologers will not only lose their name but also bring discredit to the science and to their fellow genuine Astrologers. It is high time that Legislation should intervene to prohibit such half-learned astrologers trading upon men who are already in distress. For, usually one in distress only consults an Astrologer. All these must go and Research work must be installed. That is the only way for the revival of this sacred science.

Now about this publication, I had intended to publish only the results of my Research work leaving off the elementary portions to the Readers them-selves. But many of my friends requested me to bring forth a self-sufficient and exhaustive Treatise So, I have attempted at it and placed before the public in two parts. This treatise contains very many new points not known to many so far I have tried my best to render the science mathematical. I have selected only those theories that have stood the test in all cases leaving off unproved ones. As my object is to spread these novel ideas I feel satisfied if the Readers follow my theories faithfully and adopt them in their future handling.

This treatise has two parts. The first part contains all the preliminaries and the general cannons that have to be learnt by all. I named it as Astrology made easy. It is Astrology part (I) of my Astrology books.

Second book is known as Astrology – Division Charts, Thithi, Stars, & Yogas. It is Astrology part (II) of my Astrology books.

To make it still easy to understand I wrote Astrology Lessons. In this book simple rules are narrated. It is an extension of discussions on Astrology, written in Part I and Part II.

Lastly I wrote Astrology Examples. In this book I wrote predictions and stated arguments. This shall help in understanding the subject with reasons.

Definitions and Nomenclatures of the fundamental elements are described.

Rasis and Solar Months

House no 1 is known in English Aries, In Hindi it is Mesha. It is owned by Mars. This means Mesha belongs to Mars. Mars is master of this house. Mars is known as Mangal or Kuja in Hindi language. Kindly read the following on this basis. Abbreviation used in this text is also written in the last. For Example for Mars abbreviation shall be KJ.

Pisces Venus 27° Meena JUPITER	Aries Sun 10° Mesha MARS	Taurus Moon 3° Vrishbha VENUS	Gemini Mithun MERCURY
Aquarious Kumbh SATURN			Cancer Jupiter 5° Kataka Moon
Capricorn Mars 28° Makar SATURN			Leo Simha SUN
Saggittarius Dhanush JUPITER	Scorpio Rahu / Ketu Vrishchika MARS	Libra Saturn 20° Tula VENUS	Virgo Mercury 15° KANYA MERCURY

Exaltation - Planets in Red
Name of Rashi in English & Hindi in Black
House owned by Planet in Blue.

2015/12/18 15:36

Name of Rashi in English and Hindi. Houses owned by planets.

Exaltation of planets and houses

House No.	Rasi / Solar Month	House Owner
1	Aries or Mesha	Mars / Mangal / Kj
2	Taurus or Vrishbha	Venus / Sukra / SK
3	Gemini or Mithun	Mercury / Buddha / Bd
4	Cancer or Kataka	Moon / Chandra / Ch
5	Leo or Simha	Sun / Ravi / RV
6	Virgo or Kanya	Mercury / Budha / Bd
7	Libra or Thula	Venus / Sukra / SK
8	Scorpio or Vrischika	Mars / Mangal / Kj
9	Sagittarius or Dhanush	Jupiter / Guru / Gr
10	Capricoranus or Makara	Saturn / Sani / Sn
11	Aquarius or Kumbha	Saturn / Sani / Sn
12	Pisces or Meena	Jupiter / Guru / Gr

Divide the universe by 30 degree. So 360/30 = 12 houses. These houses are owned by 9 planets and 27 stars.

Rahu / Sarpi / Rh owns Kumbha and Ketu / Shigi / Kt owns Mesha.

Birth Yogi is Lord of star of Birth Yoga Point

Duplicate Yogi is lord of the house of Yoga Point

Avayogi is sixth Disha lord from Birth Yogi

Birth Yogi Sun-Moon-Mars-Budha-Guru-Sukra-Sani-Rahu-Ketu
Ava Yogi   Sani-Budha-Ketu-Sukra-Sun- Guru-Moon-Sukra-Rahu
This means if birth yogi is Sun, then Ava yogi is Sani and if Birth yogi is moon than Ava yogi is Budha and so on.

Years (Varsha)

Reckoning of year is of several modes. West counts it from first of January. Barhaspathya – Mana is counted from the day Jupiter enters a sign. Lunar Year is counted from Chaitra Shukla Padyami and Solar year is reckoned from the time Sun enters Aries (Mesha). For Astrology Solar year and Solar months only have to be considered. There are sixty years forming a cycle. They are said to be named after the sixty sons of Narada. Here I wish to point out a special feature. No useful purpose would be served by simply narrating an element if it could not be made use of for predictions. There are books speaking on the effects of years, months, Thithi etc., but not a single effect fits in to individual horoscopes. I always value things and elements said with respect to Zodiac (Birth Chart). Unless they have a specific place in the birth chart no useful purpose will be served by merely honoring them for name sake.

It is an established fact that Kaliyuga started in the year Pramadi at the first point of Mesha. Starting with Mesha each sign has five years

Stars and Tatwa

Ashwini, Bharani, Kritika, Rohini, Mrigsira, are Prithwi / Earth Tatwa or Nitrogen element.

Aridra, Punarvasu, Pushyami, Ashlesha, Magha, Pubba are Jal / water Tatwa or hydrogen element

Uttara, Hasta, Chita, Swati, Vishakha, Anuradha are Agni or Fire Tatwa or carbon element

Jyeshta, Moola, Poorvashada, Uttarashada, Shravan are Vayu / air Tatwa or Oxygen element

Dhanishta, Shatabhisha, Poorvabhadra, Uttarabhadra, Revati are Akash Tatwa or energy

Ashwini, Punarvasu, Pushyami, Hasta, Anuradha, Shravana, Purvabhadra, Uttarabhadra are Males.

Bharani, Kritika, Rohini, Aridra, Ashlesha, Makha, Pubba, Uttara, Poorvashada, Uttarashada, Dhanishta, Revati are Females

Mrigsira, Moola, Shatabhisha are Eunuchs

Star should be noted as per Rasi chart and not as per Bhava chart or Cuspal Chart

Lord of Good Bhava is Functional Benefic. We write as FB

Lord of Dushtana or bad place is Functional Malefic. We write FM

Tatwa

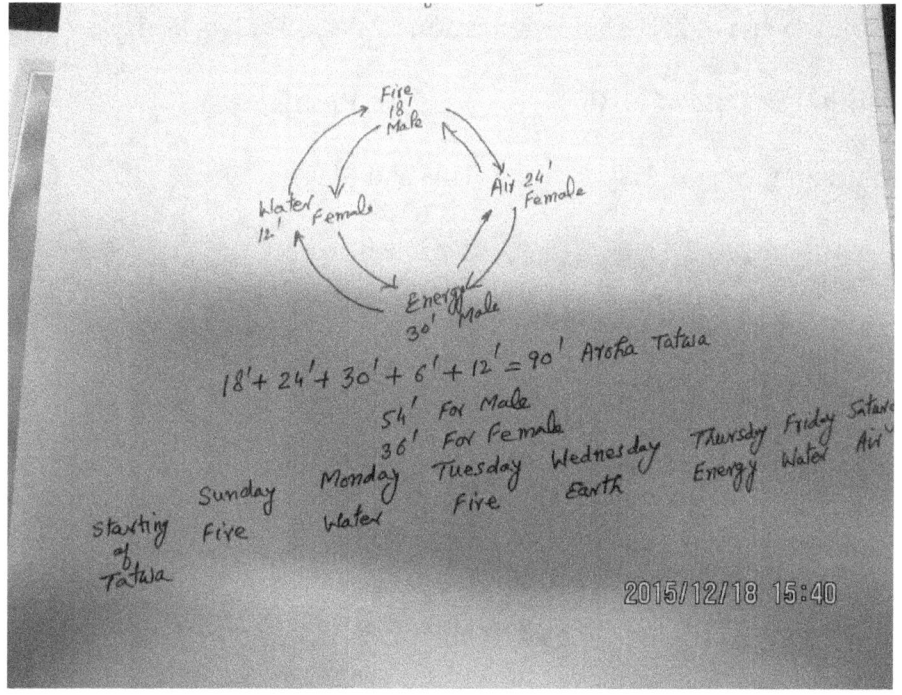

Aroha Tatwa for example for Sunday is Agni 18', Vayu 24', Aakash 30', Prithwi 6', and Jal 12'. It is total 90 Minutes. Out of 90' Male are born in 54' and Female 36'. After 90 Minute reverse cycle starts and it is known as Avaroha Cycle. Details shall be again explained.

Thithi and Zero Rasi or Dagdha Rasi or Rashi under Fire or burning Rasi

In Sanskrit Thithi means reduction and Yoga means addition. Thus the former is got by the difference and the latter by addition of the longitudes of Sun and Moon. There are 30 Thithi in all in a lunar month distributed to 360 degrees of the Zodiac. Thus each Thithi has a span of 12 degrees in the Zodiac. The first one commencing from the point of conjunction of Sun and Moon (New Moon). By name there are 14 Thithi repeating in the two halves of the month (Shukla Paksha and Krishna Paksha)

(Moon – Sun) divide by 12 = (1-9-54) – (2-4-34) = 11-5-20 or 335 Degrees 20 Minutes. Add 12 Rasis to Moon if it is less than Sun's.

Dividing 335-20 by 12 we get quotient 27 and remainder 11-20. So it is the fag end of 28th Thithi viz. Krishna Paksha Thrayodashi.

Thithi or Date of birth	Zero Rashis
Padyami means 1st Day	Tula and Makara
Dwitiya means 2nd Day	Dhanush and Meena
Tritiya means 3rd Day	Simha and Makara
Chaturthi means 4th Day	Vrishbha and Kumbha
Panchami means 5th Day	Mithun and Kanya
Shashti means 6th Day	Mesha and Simha
Saptami means 7th Day	Kataka and Dhanush
Ashtami means 8th Day	Mithun and Kanya
Navami means 9th Day	Simha and Vrischika
Dashami means 10th Day	Simha and Vrischika
Ekadashi means 11th Day	Dhanush and Meena
Dwadashi means 12th Day	Thula and Makara
Thrayodashi means 13th Day	Vrishbha and Simha
Chaturdashi means 14th Day	Meena, Mithun, Kanya and Dhanush
Full Moon	None
New Moon	None

To Locate Yoga Point

Add Rasis 3-3-20 (Sign – Degree – Minute) to the sum total of Longitude of Sun and Longitude of Moon. You get exact location of Yoga Point. Name of each Yoga is Vishkhamaba, Preeti, Ayushman etc. Kindly read the following

Star - Planet - Yoga - Effect

1.0 Pushyami– Saturn - Vishkhamba - He will win over and over come others. Wealthy. He is blessed with cattle and wealth.

2.0 Ashlesha– Budha - Preeti -- He is loved by all, attracted by women

3.0 Makha– Ketu - Ayushman – Has good longevity and health

4.0 Pubba– Sukra – Saubhagya – Blessed with Happiness and comforts

5.0 Uttara– Ravi – Shobhana – Lascivious, Sex minded

6.0 Hasta– Chandra – Atiganda – Murderer, Meeting many accidents in life

7.0 Chitta– Kuja – Sukarma – Doing good and noble acts

8.0 Swathi– Rahu – Dhriti – Indulging in others money and women

9.0 Vishakha – Guru – Shoola – Angry and quarrelsome

10.0 Anuradha – Saturn – Ganda – Bad Character Person

11.0 Jyeshta – Budha – Vradhi – Improving day by day and intelligent

12.0 Moola – Ketu –Dhruva – Fixity of mind and wealthy

13.0 Poorvashada – Sukra – Vyagata – cruel minded person

14.0 Uttarashada – Ravi – Harshna – Always merry going and intelligent

15.0 Shravan – Moon – Vajra – Wealthy and Lascivious

16.0 Dhanishta – Mars – Siddhi – He will have several attainments and protector of others

17.0 Shatabhisha – Rahu – Vyathipatha – Unreliable

18.0 Poorvabhadra – Guru – Variyam – Bad Character and Lascivious

19.0 Uttarabhadra – Saturn – Parigha – Wealthy and Quarrelsome

20.0 Revati – Budha – Shivam – Revered by Kings, Cool, calm, well versed in Shastra's, wealthy

21.0 Ashwini – Ketu – Siddham – Good natured and interest in religious rites

22.0 Bharani – Sukra – Sadhyam – Good Manners

23.0 Kritika – Ravi – Shubham – Wealthy, lustrous, fair and unhealthy

24.0 Rohini – Chandra – Shubham – Wavering Mind, Good Mannered, Talkative, Impulsive

25.0 Mrigsira – Mars – Brahma – Highly confidential, high aspirations, has capacity to judge correctly.

26.0 Aridra – Rahu – Indira or Mahendram – Wealthy, learned and helping nature

27.0 Punarvasu – Guru – Vaidriti – Cunning, blaming nature and wealthy.

Ritu (Two Months)

A Ritu consists of two Lunar Months. Thus there are 6 Ritu in a year.

Ritu	Months and Stars
Basanth	Chaitra and Vaisakha (Chitta and Vishakha)
Grishma	Jyeshta and Ashada (Jyeshta and Poorvashada)
Varsha	Shravana and Bhadrapada (Shravana and Poorvabhadra)
Sharad	Ashvija and Kartika (Ashwini and Kritika)
Hemant	Margshira and Pushya (Mrigsira and Pushyami)
Sishira	Magha and Phalguna (Makha and Uttar Phalguni)

These Lunar Months takes their names after the star on the full Moon day of the Month. The Stars on the full Moon days of the respective months are written in English

Paksha (Half a Lunar month)

Shukla Paksha and Krishna Paksha rules a Lunar Month. From the time of separation of Sun and Moon to the point of direct opposition it is named as Shukla Paksha. From opposition to Conjunction it is Krishna Paksha.

Vara

There are seven week days from Sunday to Saturday. As per western calendar these days rule from Midnight to Midnight while as per Hindu and Astrological purposes the week day always commences from Sun Rise. The week day takes its name after the Hora at Sun rise.

Hora (One Hour)

Each Hora rules an hour. It looks as if the word Hour is derived from Sanskrit word Hora. There are seven Hora on the basis of seven Planets Sun etc. leaving of Rahu and Ketu who are mere Chaya grahas (Shadow Planets). These Horas have a particular order. Sun –Venus – Mercury - Moon – Saturn – Jupiter – Mars. In this order they complete their full cycle a day covering 21 hours as each Hora rules an hour. The first Hora takes the name of that week day. Thus after 3 cycles in a day there remain 3 Horas to complete the day. The fourth Hora starts the next week day. Thus you find that observing the order of Horas the fourth represents the next week day. From Sun Hora the fourth Hora is Moon. So after Sunday you have Monday. From Moon Hora the fourth Hora is Mars, so Tuesday. Fourth Hora from Mars Hora is Mercury hence Wednesday. Fourth Hora from Mercury Hora is Jupiter Hora, hence Thursday and So on. Please note that Hora plays a very important part in deciding the time of daily predictions which will be treated later.

Graha Samyam (Planetary State)

While reading the effects of a Major Lord (Udu Dashanath) it would be also helpful to know this additional character obtaining from its state. Even here there are different versions but the one I state below may be taken as authentic since it has proved to be true in all cases. There are 27 states in all

Method of Finding the State:-

Count the number of Rashis from Mesha to Lagna and again count the number of Rashis from Lagna to the Planet (Disha Lord). Add the two and multiply the sum by twice the number of years allotted to the planet under Udu Disha. Divide this product by 27 and consider the remainder. This represents its state.

Do not take literal meaning of these effects. Interpret them to suit the situations.

Snanam (Bathing)

Vastra Dharanam: Honored by King and government; acquisition of money, clothes, and precious metals, scents, perfumes and Ornaments, Possessing good strength

Vibhuthi or Gandhalankaram: Fame or Scented decoration, State honors, happiness and Mirth; ability in work

Shiva lingam or Pooja Yatnam (Attempt to Worship); Money through lands; acquisition of vehicles, happy living, revered by kinsmen

Panchaksharijapam (Chanting of five Letters): Acquisition of lands and Money; trouble by Government; loss of Money

Shiva Pooja (Worship of Ishwar): Association even with wicked men, Love of people; monetary gains

Note

1.0 Stars to be noted as per Rasi chart and not as per Bhava Chart or Cuspal Chart.

Longitudinal span of a Rasi is 30 Degrees. Distributing these 30 degrees to 9 Padas of a Rasi, we get 3 degree 20 Minute as the longitudinal span of a star pada. As such 4 pada of a star means full star. It constitutes 13 degree 20 Minute span of a star in a Rasi of 30 degree.

There are 9 Padas in a Rasi. This means in a Rasi there shall be two full star and 9th Pada shall be one pada of one star and so on. 12 Rasis means 12 X 9 Padas=108 Padas.

CHAPTER II   COMMODITIES CHART & MAIN FEATURES

COMMODITIES CHART

12	1	2	3
11			4
10			5
9	8	7	6

1   First house represents Cotton, Dry crop woolen, Ragi, Fabric, Oil, Jars , Leather, Beans, Barley, Wheat

2 2nd house or No.2 represents Clothes, Wheat, Buffalo, Cow, Rice, Flowers, and Barley

3 Autumnal, Crops, Creepers, Lilly Root, Cotton

4   Kodrawa, Bent Grass Roots, Leaves Etc., Barks, Plantain, Fruits

5   Food Grains, Juice, Lion's Skin and of like animals, Jaggery

6. Common Flax, Barley, Horse-Gram, Wheat, Kidney Beans, Nishpava

7   Black Gram, Wheat, Mustard Seed

8   Sugar Cane, Wet Grains, Metals, Woolen Fabrics,

9 Horses, Cloth, Weapons, Weeds, Roots, Salt, Gingerly

10 Trees, Bushes, Wet Grains, Sugar cane, Gold, Black led, Black Gram, Wheat, Mustard Seed

11 Water Fruits, Flowers, Gems, Articles of various shapes

12 Pearls, Shells, Diamonds, Various oils

Remember The articles of the sign perish when:-

Guru transits in Gochara: - 2-4-5-7-9-10 or 11 Rasi from it

Budha transits in Gochara   2-5-8-10 or 11 Rasi from it

Sukra transits in Gochara   6 or 7 Rasi from it

In other positions the articles flourish.

2) If the above benefic planets transiting favorable position are also powerful the articles can be had easily and for a moderate price.

3) a) If malefic transits 3-6-10-11 (Upachaya) from any sign the articles of that sign flourish. In other positions they perish.

b) If powerful malefic transit Apachaya positions (other than 3-6-10-11) the articles sell dear and become scare.

Now I give in detail their characteristics.

(A) MESHA (ARIES)

Forest, Quadruped ,Pristodaya, Movable, Odd, Dwara-Rasi, Mineral, Fierce, Male, East, Red, Head, Night, Goat, Kshatriya Agni - Tatwa, Agriculture, Bilious – Fever, Liver – Disease, England ,Germany, Peru, Syria.

(B) VRISHABHA (TAURUS)

Meadows, Wetlands, Quadruped, Pristodaya, Fixed, Even, Bahia Rasi, Vegetable, Soft, Female, South, White, Face, Night, Bullock , Vaisya, Prithwi- Tatwa, Dairy – Farming, Ireland, Persia, Poland, Cypress, Asia-, Minor, White-, Russia.

(C) MITHUNA (GEMINI)

Bed room, First Half, Biped, Ubhayodaya, Common, Odd, Garbha –Rasi, Human – Rasi, Fierce, Male, West, Green, Village, Neck, Night, Sudra, Vayu – Tatwa, Low professionals, Catarrh, Asthma, Colic pain, United states of America, Belgium ,North Africa, Wales, New Orleans (27^0), London (18^0).

(D) KATAKA (CANCER)

Chasm with water Channel, Watery – sign, Pristodaya, Movable, Even Dwara, - Rasi, Mineral, Soft, Female, North, Heart, Night, White, Brahmin, Jala – Tatwa, Watery Animals, Smithy work, Insanity, Windy disease, Tastelessness, China, Holland, Canada, New Zealand, New York (14^0).

(E) SIMHA (LEO)

Mountain, Quadruped, Shirodaya, Fixed, Odd, Bahia-Rasi, Vegetable, Fierce, Male, East, Stomach, Day, East, Lion, Kshatriya, Tejo –Tatwa, Barbers - Profession, Fever, Boils, France, Italy, Sicily, Rumania, Bohemia.

(F) KANYA (VIRGO)

Town Land, full of water, and corn, First half Biped, Shirodaya, Common, Even, Garbha, Rasi, Human, Rasi, Soft, Female, South, Variegated colors, Waist, Day, Maiden, Vaisya, Prithwi Tatwa, Boat, driving, Venereal diseases, Turkey, Greece, Switzerland, West Indies, Babylonia.

(G) THULA (LIBRA)

House of Vaisya, First Half, Biped, Shirodaya, Movable, Odd, Dwara – Rasi, Mineral, Fierce, Male, West, Dark – color, Lower – Abdomen, Day, Weighing – Balance, Sudra, Vayu Tatwa, Human – Rasi, Trade, Brain – fever, Typhoid, Austria, Argentina, Japan, Gujrat, Tibet, Burma.

(H) VRISCHIKA (SCORPIO)

Hole or Cavity, Reptile, centipede, Shirodaya, fixed, Even, Bahia Rasi, Vegetable, Soft, Female, North, Brown, Well, Sex organs, Day, Scorpion, Brahmin, Jala – Tatwa, Hunters- Profession, Diseases of Spleen, Jaundice and Sprue, Brazil, Norway, Transvaal, Morocco, Bombay (2^0).

(I) DHANUS (SAGGITARIUS)

King's residence, First Half Biped, Latter Half, Quadruped, Pristodaya, Common, Odd, Garbha, Rasi, Fierce, Male, East, Brown-color, Forest- Garden ,Thighs, Night, Arrow, Tejo- Tatwa, Goldsmith, Arabia, Australia, Hungary, Spain.

(J) MAKARA (CAPRICORNUS)

Water abounding forest, Latter half, watery animals, First half, quadruped, Pristodaya, Movable, Even, Dwara Rasi, Mineral, Soft, Female, South, Mixed colors, Lifeless, river, Knees, Night, Crocodile, Vaisya, Prithwi Tatwa, Laundry work, Stomach ache, Want of Appetite, Aberration of mind, Albania, India, Bulgaria, Mexico.

(K) KUMBHA (ACQUARIOUS)

Potters place, Shirodaya, Biped ,Fixed, Odd, Bahia Rasi, Vegetable, Fierce, Male, West, Tank and pond, Pot, Sudra, Vayu Tatwa, Buttocks, Potters professions, Cough, Fever, Consumption, Abyssinia, Prussia, Tartary.

(L) MEENA (PISCES)

Watery place, Sea, Watery animals, Ubhayodaya, Common, Even, Garbha Rasi, Humanrasi, Soft Female, North, Blue, Feet, Night, Fish, Brahmin, Jala Tatwa, Fishing, and other low pursuits, Watery diseases such as Ascitis (Jalo-dara), Portugal, Normandy, Galicia, Egypt.

(2) BHAVAS (HOUSE)

1   Six Houses from Lagna to Sixth bhava represent the right side limbs of Kalapurusha while the other six from twelfth to seventh in reverse order denote left limbs,

2   1-4-7-10 houses from Lagna / ascendant are Kendra (quadrants)

2-5-8-11 houses from Lagna are Panaphara (succedents)

3-6-9-12 houses from Lagna are Apoklimas (cadent)

5-9 houses from Lagna are Trikonas (trines)

3-6-8-12 houses from Lagna are Dushtana

3   Second Bhava represents classic and family education and Fourth Bhava represents higher education of the times.

4   1-5-9 Houses are termed Dharma houses charitable

2-6-10 houses are termed as Artha houses monetary

3-7-11 houses are termed as Kamya houses desires

4-8-12 houses are termed as Moksha houses blissful

5.   2-4 Bhavas connote paternal property and goods.

10th Bhava connotes Self Earned with exertion

11th Bhava connotes Self Earned without exertion

The Connotations of Bhavas are detailed below.

(A) Lagna

Body strength and constitution, complexion, Head, Appearance, Beauty Department, Name, Fame, Piety, Conduct, Perception, Happiness, Current birth, birth place, longevity.

(B) Second Bhava

Family, Money, Right Eye, Eye Sight, Speech, Expression and Eloquence, Authorship, Hands, Meals, Food and drink, face , learning, letter or document, belief in sacred tradition.

Notes: - Family: - Immediately the child is born it has first contact only with the mother. Then its family is mother only. Some days later it recognizes father. Then its family is Father and Mother. After some time it recognizes all those in the house. Then its family is all the members of the house. After marriage the partner gets in and after fatherhood the children and then grand – children. Like this the word Family is flexible. While reading these effects confine only to the relatives that come to play part at different ages.

Learning: - Here it confines to learnings in traditional knowledge like Scripture, Sastras, Vedas as Astrology, Ayurveda etc.

(C) Third Bhava

Younger brothers and sisters (after co born), Bravery and Prowess, Ear, Servants, Minor, Debts, Chest, Partition of property

(D) FOURTH BHAVA

Mother Relations and Friends Higher Education House and Lands Equipages Comfort and Easy going life Treasure Trove Trade and Commerce Water and Bathing Wells and Tanks, Conjugal Life and Sexual enjoyment, (Specially in Female Horoscope) Heart, Cow, perfume, clothes, ornaments, happiness, bridge and river.

(E) FIFTH BHAVA

Issues, Intelligence, Intellect, Spiritual knowledge and Religion, Past Karma, Yoga Practice and Mantras, Intuitive power, Abstract of Mind, Superiors, Royal favor and patronage, Brain, Lower abdomen and Uterus, Minister, Tax or Toll, Ataman, knowledge of future, Belly, Sruti (Vedic Knowledge), Smirti (Traditional Law), Departed Spirits (Asu), Pregnancy.

(F) SIXTH BHAVA

Enemies, Debt, Disease, Theft and Loss, Hatred, Enmity and Quarrel among relations, Fasting, Bodily tortures, Venereal diseases,

Criminal Prosecution and litigation, Imprisonment and penal Servitude, Wounds or Carbuncles, Domestic Misery and disgrace, Travel from place to place, Transfers, Hip.

(G) SEVENTH BHAVA

Husband or Wife, Sensual desire and enjoyment, Sexual organs, Son-in-law, Desires, Increase of family burden, Issues to the second or subsequent wives, Trade, Death, Marriages, Travel Groins, passion, Gambling, Public Open Opponent, Litigation, Kidneys.

Why issues to subsequent wife is given is not known?

(H) EIGHT BHAVA

Duration of Life, Death, Destruction, Loss, Sudden Death and Murder, Death in Battlefield, Fall from a tree or a high place, constitutional and continued Sickness, Disappointments and Failures, Prosecution, Persecution and Disgrace, Trouble, Imprisonment and penal servitude Hanging on the gallows, Suicide, Defeat, War, Epidemics Emigration to foreign lands, Distant and sudden transfer, private parts, Widowhood, will, Legacy.

This is both a house of Longevity and Death. Its lord being powerful elongates life while a malefic in the Bhava shortens. Usually all the effects of this Bhava arise suddenly without one's previous knowledge. Sickness read from sixth house may be temporary and curable while that of the eight Bhava is longstanding incurable and even hereditary.

(I) NINTH BHAVA

Deeds of Virtue, Dharma, Father, Preceptor, Wealth, Superior, Paternal property, Charity, Initiation into the sacred Vestals of Religion, Astamaha Siddhi and yoga practices, Spiritual knowledge, Renunciation of worldly ways and living as a Recluse in Caves and Forests, Medicine, Alchemy, etc. Performance of Yoga's and Holy Sacrifices Trade, Acquisition of Wealth, Faith in God, Abolition of sins done by fore-fathers, Fame, Thighs, Previous birth, Luck, Worship, Penance Virtue or Religious merit, A good or Virtuous Act, Grandson, Distant travel or Sea Voyage.

Notes:-

1- GRANDSON. This house is 4th from 5th (House of Issue).

2- TRAVEL: Travels are read out from various house viz, 3-6-7-8-9-12. In each case there is a difference. Travel of 3rd house is not good; so also of 6th. Travels of 8th arise suddenly mostly for bad and of 12 to far distant places. But that caused by 7th is happy sojourn for short time-often going to and for pleasure or business that of the Ninth is the best that earns one name, fame and prosperity and foreign tour.

(J) TENTH BHAVA

Avocation, Karma Sthana, Fame, Knees, Commerce and Trade Service, Rank or Position, Honor, Livelihood, Sky, Inclination, Gait, Command Superior officer.

Notes - **KARMASTHANA** - Means either doing some work or do the obsequies of the departed.

(K) ELEVENTH BHAVA

Earning and Gain, Acquisition, Elder brother or sister, Enjoyment of many women, Chamaran, Chowries, Salutes and all honors and State honors, Treasure- Trove, Fulfillment (Siddhi) and Attainment Commendation, Left, Ear, Hearing of some pleasing and delightful news, Friends, Second wife or companion.

(L) TWELTH BHAVA

Loss, Bad deeds, Expenditure, Distant Travel, Sleep and Bed Comforts, Dispute, Litigation, Prosecution, Punishment, Imprisonment, Salvation and liberation of the soul (Mukti) Feet, Left, eye, Declining Secret, opponent, Hospital Suicide.

Note: - The only good aspect of this house is Final Bliss. If Ketu (the Karaka for Bliss) be in 12th Bhava powerfully and auspiciously situated and if he dies in Ketu Disha he will have no more births.

(3) PLANETS

1-   Highest Exaltation Degrees of Sun Onwards in order are -

   10-3-28-15-5-27 and 20 respectively.

2-   Sun, Rahu and Jupiter cause multiple sons.

   Moon gives only one Son. (May give more daughters).

   Kuja, Budha, Sani Cause Adoptive Sons.

3-   Sun, Rahu, and Guru are good for progeny

   Sun is good for partner's welfare.

   Sani is good for longevity and younger co-born.

   Guru, Sani, Ketu are good for Ethical and moral code.

   Rahu makes one bereft of Rituals.

   Moon and Sukra make on showy.

4-   Types of Sexual unions are with:-

   Sun   :   Family Ladies.

   Moon, Guru, Sukra: Wife.

   Kuja: One with bodily deformities.

Budha: Artisan.

Sani: Over-aged woman; Sickly and a Saint (Thapasvini)

Rahu: A widow or a divorced lady.

5-   Types of Buildings.

Sun: Thatched Hut.

Moon and Sukra: Terraced.

Kuja and Ketu: Mud or Bricks.

Budha: Artistic.

Guru: Wooden

Sani and Rahu: Stoneware.

6-   Aspects

All planets aspect the seventh house. In addition to this the following are the special aspects.

Kuja: 4 and 8

Guru: 5 and 9

Sani, Rahu, Ketu: 3 and 10.

7- Sun represents the Human Soul

Moon represents Mind

Mars Brute Force and Strength

Budha Speech and worldly knowledge

Guru Spiritual knowledge and Intuitive Skills

Venus Cupidity, Desire, and worldliness

Saturn Sorrow and Labor

8 Sun and Moon are kingly in the sense that Sun is King and Moon is Queen, Mars is the Commander- General, Mercury is the prince or Heir Apparent, Jupiter and Venus are Ministers and Saturn the Servant.

9- Castes: According to Varahamihira:

Guru and Sukra are Brahmins. Sun and Kuja are Kshatriyas. Moon a Vysya, Budha a Sudra and Sani a Chandala.

But Sarvartha Chintamani says:-

Guru and Sukra are Brahmins. Sun and Kuja are Kshatriyas. Moon and Buddha are Vaisyas or Trading

Class, Saturn the Shudra class, Rahu a Pariah, Ketu other lower castes.

The latter classification appears to be more appropriate. A combination of Rahu and Sukra gives Christianity.

Sukra and Sani: Mohammedanism

Guru and Sukra   Buddhism.

10-   Planetary Gunas (Temper):

Sun, Moon, Guru are Satwics (good temper) Kuja, Budha, and Sukra are Rajasic (Passionate temper).

Sani, Rahu, Ketu are Thamasic (Lethargic).

11-   Planetary Looks: (Jinendramala says).

Moon and Guru look straight. Kuja looks aside sun looks above. Budha and Sukra Look down. Sani and Rahu Look obliquely.

12   Planetary Tastes – Sarvartha Chintamani says

Sun governs hot and pungent taste, moon Salty, Mars bitter, Budha all tastes (Shad Rasa); Guru Sweet; Sukra sour; and Shani bitter tastes (Astringent).

13- Planetary Vegetables and Fruits.

Sun represents chillies, Radish, etc.

Moon all cold substances and tender fruits and Vegetables.

Kuja all substances of a fleshy nature. Groundnut, Pulses or Dhall, Grains etc.

Budha Plantains, Brinda's, Ladies finger, Beetroot etc.

Guru all native roots and bulbs, Pumpkin Plantains, etc.

Sukra all exotic Vegetables, Potatoes, Cabbages etc.

Sani Bitter gourd, Onions, Drumstick, Betel leaves, Tobacco and Greens.

Rahu and Ketu - Snake gourd, Flavor substances, Garlic, spices, etc.

14-   Planetary Trees -

Sun represents Mountain trees, Moon represents Coconut tree, Kuja- Ginger, Millet, Dhal, Bengal Gram, and thorny trees. Budha and Guru - plantains and Wet crops. Sukra / Venus -   fruit trees, flower trees and Creeping plants. Sani and Rahu -   Palmyra thorny and poisonous trees.

15-   Planetary places (Varahamihira states)

Sun presides over place of Worship, Moon wells, Kuja fireplace, and Budha playgrounds, Guru Store-room, Sukra Bedroom, Sani, Rahu and Ketu where sweepings are gathered.

SARVARTHA CHINTAMANI IS MORE CLEAR ON THIS POINT

1. Sun rules over place of worship and temple.
2. Moon Bathrooms, Tanks, Wells and other Watery surfaces
3. Kuja Fireplace, Kitchen engines and armories. Budha Playgrounds.
4. Guru Treasury and places where money and Jeweled are deposited.
5. Sukra Bedroom and Drawing Room
6. Sani Hospitals and Medical.

7. Rahu and Ketu where serpents and Reptiles hid themselves.

Rahu represents snakes and like reptiles having poison in the mouth.

Ketu represents scorpion and like having poison in the tail.

16- Planetary Sciences and Shaka (Veda)

The Author of Laghujataka says-

Sun and Budha Combined denote Vedic literature in general. Guru is lord of Rigveda, Sukra- Yajurveda, Kuja- Sam Veda, and Budha - Atharvana Veda.

The Shaka of a Brahmin may be determined by the strongest planet at birth (even if it does not aspect Lagna).

Sun represents Veda, Medicine, and Alchemy.

Moon- Aesthetical Sciences, Music and Textile works.

Kuja- Culinary Sciences, War History and Engineering

Budha- Pure mathematics, Astronomy, Accounts, Drawing, and computer science

Guru- Philosophy, Thought- Reading, Yoga philosophy, etc.

Sukra- Naval Science, Law, Literature, Poetry Music, Foreign Literature, Logic, Grammar, etc.

Rahu and Ketu- Drama, Jugglery, Prestidigitation (one who plays slight- of - hand), Buffoonery and the like.

17- Planetary Nature Physical and Physiological, (Sarvartha Chintamani).

Sun rules over bones, Moon - blood, Kuja - Marrow of the bone. Budha- Skin, Guru - Brain, Sukra- Semen or Vital fluid, Sani - Muscular and Nervous system.

Notes: The diseases of person or his health depends upon the strength of the planets and the organic matter them government.

The Natural Diseases ascribed to planets are:

a) Sun- Fever dominated by Liver and heat, Eye- Disease, Dental trouble, Neuralgia (Nervous Pain).

b) Moon- Sleeping Disease. Drowsiness, Diseases of the Lungs (Asthma and consumption), Diarrhea, Lunacy, Phlegmatic Complaints, Tastelessness, Debility, Indigestion , Jaundice,

Impurity of Blood, Bala Graha Peeda, Danger from water, Cholera and watery Diseases.

c) Kuja - Bilious Fever Injury in the marrow of the Bone, Heat complaints, Small-pox Ulcers.

d) Budha - Mental Disease, Skin Diseases - Anemia, Liver complaints.

e) Guru - Appendicitis, phlegmatic Diseases, Anemia, Liver complaints. Ear trouble.

f) Sukra- Anemia, Liver and Bilious complaints, Jaundice, Seminal disorder, Urinary Diseases, Disease in the general organ. Trouble in or by co - habitation. Exudation of semen, Loss of bodily splendor by intercourse with women.

g) Sani- Windy and phlegmatic complaints, Belly- Ache, Paralysis and Rheumatism, Dyspepsia (Indigestion) Body deformity, cerebral disorder front and large and larger part of Brain.

h) Rahu and Ketu: Epidemics. Hysteria, Venomous and Poisonous Complaints, Epilepsy, Dyspepsia, Carbuncle, Cancer; Dropsy, Eczema and other Skin Diseases, Leukoderma and Leprosy.

Rahu specially causes Palpitation of the Heart Leprosy, Aberration of the mind, and Danger from poison, Pain in the legs, Trouble from Goblins and Serpents.

Ketu specially causes all poisonous diseases especially consumption, Scorpion bite and Typhoid.

18- Planetary Garments (Varahamihira).

Sun's garment is of thick thread, Moon-Fine and new, Kuja-Rough and partly burnt (singed). Buddha- Wet Cloth, Guru-Ordinary and somewhat used, Sukra- Strong texture, Sani - Torn clothes, Rahu and Ketu- Rags.

19- Planetary Colors (Varahamihira).

Sun is of Red Dark-brown color and presides over copper color. Moon is White and presides over Pearls and White color. Kuja is both Red and White presiding over Blood Red color. Budha is Green. Guru is yellow like Molten Gold. Sukra is a mixture of various colors (vibgyor) and presides over light Blue or ultraduraine color. Sani and Rahu are Black. Ketu is Dark- Red.

20- Planetary Metals.

Sun presides over copper. Moon on Gems, Pearls, Corals. Kuja Gold, Budha- Brass, Guru Gold and Cat's eye. Sukra - Silver, Pearls, Diamond. Sani - Iron and Mercury, Quick- Silver and Emerald.

Notes: These are useful in ascertaining the color of the stolen articles and in determining the color of the flowers to be used in the worship of planets. Even in the administration of medicine for strong disease the above may also be found useful. For example, for diseases indicated by sun in his period medicines with the chemical combination of Copper may be administered. For moon oxides of Pearls of Corals. For Kuja Auriferous Compounds. For Sukra oxides of Silver, Pearl, or white Poison. For Sani Mercurial or Ferric compounds. For Rahu like Sani and for Ketu like Kuja.

21-   Planetary Directions.

Sun-East, Moon- North West, Kuja - South, Budha- North. Guru North East, Sukra - south East, Sani West,   Rahu and Ketu   South West.

Notes: These are useful in deterring the position of the Delivery Room, the direction of the escape of thieves, the direction of travel, transfer, etc.

22-   Planetary Limbs.

Sun - Head, Moon - Face, Kuja- Chest, Budha - Hips, Guru - Belly. Sukra - Pelvis and sexual organs. Sani - Thighs, Rahu- Two hands. Ketu Two legs.

Notes: These are useful both for Horary and Horoscopy. The diseases of the organ may be located.

23- Planetary Geometrical shape (Jinendramala)

Sun is Quadrangular. Moon small circle, Kuja small drum (Damaru) shaped like an Hour glass, Budha- Triangular, Guru - Elliptical, Sukra - Octagonal, Sani- Shape of a Window, Rahu a line, Ketu Flag.

Notes: In preparing lockets, pendent, rings the shape of one's favorable planet will be propitious.

24- Planetary Legs,

Budha, Guru & Sukra are Bipeds.

Sun, Kuja, & Sani are Quadiapeds.

Moon and Rahu are centipedes.

25- Planetary Distances.

Sun and Budha show 8 yojanas, Moon 1 yojna and Kuja 7, Guru-9, Sukra-16, Sani- 20 and Rahu-20 yojna.

Notes: The Distances of Rasis follow that of their Lords. These are useful in determining the distance at which the thief is, the place to be transferred is etc. Yojanas is about 9 miles.

26-   Planetary Nature of Birth and their waking periods.

Budha, Guru Sukra and Rahu are Shirodaya Planets that rise with their head. They are strong during day time. Sun, Kuja, Budha and Sani are Pristodaya planets that rise first with their feet. They are powerful during night.

Moon and Ketu are Ubhayodaya planets that are powerful both day and night.

Notes: These are helpful to find out the nature of birth as to whether the child coming out of womb shows its heads, or feet or buttocks first.

27-   Planetary Periods.

Sani and Rahu have 1 year, Sun 6 months (Ayana), Budha 2 months (Ritu), Guru 1 month (Masa), Sukra 15 days (Paksha), Kuja 1 day (Dina) and Moon 48 minutes (Muhurthem).

Note: The period of the signs are those of their Lords. This is useful in Horary to gauge the time of occurrence.

28.   Dhatu, Moola and Jeeva Planets.

Moon, Kuja, Sani and Rahu are Dhatu Planets (mineral). Sun and Sukra are Moola planets (vegetable). Budha and Guru are Jeeva planets (Living beings).

29- Planetary Stature.

Moon - Kuja -Sani   are   short

Budha - Guru – Rahu   are   tall

Sun and Sukra are of normal height

30- Planetary Deities (Saravali says):

Sun is Agni (Fire God)   Moon is Varuna (Rain God),

Kuja is Subramanian,   Budha is Vishnu,

Guru is Indra, Sukra is Indrani   &   Sani is Brahma.

This does not seem to fit in well. I therefore restate other's opinion   as that seems to be more fitting.

Sun denotes Sadashiv and Shaivism.

A combination of Sun and Sukra is Zoroastrianism (Ancient Religion founded or refunded).

Moon: Stands for Ganapathy and Shakti and Sree Vidya Upasana and Epicurean Philosophy (luxury)

Kuja: god Subramanian, Ganapathy, Veeran and War God.

Budha: Vishnu, Vaishnavism, Dualistic Philosophy (Dwaita).

Guru: Brahmana, Buddhist and Monistic Schools of Philosophy.

Sukra: Female Deities Sree Vidya Upasana.

Sani: Minor Cruel Deities such as Ayanar, Sasthan, Satan. Athriman Yama and other secular Sectarian Religious minor Deities.

Rahu and Ketu: Rakshasa and Blood - thirsty Deities. A combination of Budha and Kuja or Budha and Sani represents Hanuman. Rahu, Kuja and Sani combined causes stone worship, Fetishism (Charms) and low form of Religion.

(2)   Worship of Deities ruled by the affected planets during their periods will mitigate their evil effects.

(3)   Planetary Elements (Varahamihira)

Sun and Kuja Indicate Tejo / Agni Tatwa (fire / carbon)

Moon and Sukra – Jala Tatwa   (water or Hydrogen),

Budha   Prithwi Tatwa   (Earth or Protein or Nitrogen),

Guru   Akash Tatwa (Ether / Energy).

Sani-   Vayu Tatwa   (Air / Oxygen).

Notes: These planets also indicate some abstract physical principles and are useful to create Amino Acids used in DNA or RNA.

Sun represents Heat, Light and physical Evolution. Moon denotes Humors and Mental Evolution. Mars denotes Physical Force and Kinetic Energy. Budha represents speed and mathematical proportions. Guru gives intellectual Evolution, Universal Harmony and Appreciative Knowledge. Sukra controls Space, Electricity and Emotion. Sani rules over Time and Life Principles. Rahu and Ketu denote Destructive forces such as Dissolution, Segregation, Dis-integration, etc.

32- Planetary Season, (Varahamihira) Roughly

Planet	Ritu	Hindu Month	English Month
Sukra	Vasanta	Chaitra / Vaisakha	Mar/ April
Sun & Mars	Grishma	Jyeshta / Ashada	May/June
Moon	Varsha	Shravana / Bhadrapada	July /Aug.
Budha	Sharath	Aswija / Kartika	Sept / Oct.
Guru	Hemant	Margshira /Pushya	Nov/ Dec.
Sani	Sishira	Magha / Phalguna	Jan/ Feb.

Notes: A Planet in Lagna or Lord of Drekkana gives out the season at Birth Time.

33- Walking, creeping or Flying (Jinendramala)

Sun, Kuja, Guru and Sukra   are   walking planets.

Mon and Rahu   are   creeping planets.

Budha   is   a   flying   planet.

Sani   is   a   limping   planet.

34- Planetary Senses (Panchendriyas) Jinendramala

Sani presides over sense of Touch (Twak)

Budha - Touch and Taste (Twak and Jihva)

 Mars – Touch, taste and sight (Twak, Jihva and Chakschu)

 Sukra - Touch, Taste, Sight and smell (Twak, Jihva, Chakschu, and Ghrana).

Guru - Touch, Taste, Sight, smell and hearing (Twak, Jihva, Chakschu, Ghrana, Srothra)

Notes:

As per this Sani presides over plants as they have only the sense of touch. Budha over conches, cowry, Oyster, snails and the like possessed of the two senses of touch & taste. Kuja is an ant, a louse, a fly and the like possessed of senses of touch, taste, sight. Sukra is a wasp, a beetle, butterfly and the like possessed of four senses of touch, taste, sight and smell. Guru is a Deva, a man, an animal, a bird and the like possessing all the five senses.

These points are very useful in Horary. For the sake of convenience of predication the living creatures are divided into 4 groups   1) those that walk   2) that Fly,   3) that Creep and 4) that live in water.

Rahu and Ketu denote Venomous Reptiles and Snakes.

35- Planetary Grains.

Sun's Grain is Wheat, Moon-Rice. Kuja- Tur Dhal. Budha- Green Gram, Guru - Bengal Gram. , Sukra -Dolichos Lablab or Cow Gram (Avare). Sani Sesame. Rahu Black Gram. Ketu- Horse Gram.

36- Planetary Countries

Sun- Kalinga, Moon- Yavana, Kuja- Avanti, Budha- Magadha, Guru- Sindhu, Sukra- Keekata, Sani Saurastra, Rahu Ambar.

37- Planetary Stones and Gems.

Sun Ruby, Moon pure spotless Pearl, Mars – Coral, Budha Emerald, Guru Topaz, Sukra Diamond. Sani stainless sapphire, Rahu Agate, and Ketu Lapis lazuli or Turquois.

In Sanskrit they are, Sun Manikyam, Moon Muktha, Mars Vidrum, Budha Marakatham, Guru Pushparagam, Sukra Vajram, Sani Neelam, Rahu Gomedhika, and Ketu Vaidhurya.

38- Planets & Numbers

Planets	Number	Thithi	Letters
Sun	1-4	1-11	Avarga
Moon	2-7	2-2	Yavarga
Kuja	9	6	Kavarga
Budha	5	7-	Tavarga
Guru	3	3-8-13	thavarga
Sukra	6	4-9-14	Chavara
Sani	8	5-10-15-30	Pavarga

The following paragraphs contain consolidated list of all the Karakatwas (characteristic) of planets a valuable collection from rare books and by research· The readers will find very useful information not known so far· To predict meticulous details as you find in Nadi Reading one should master all those Karakatwas and have them at his finger's end to apply immediately at the first glance of a Chart. **By clubbing the Karakatwas of the planet, its Rasi, its Bhava and its Star** one could very definitely arrive at proper judgment. The more you master these details the more efficient you become.

Planets also represent

(A) SUN

East, Father, Soul (Atman), Head, Right eye, Limited Hairs, Idiosyncrasies, Medicine, Neuralgia and Head-Ache, Fever, King, Service under a Ruler or Sovereign (Government), Saivism, Sadashiv, Sun, God, Alchemy, Veda, Wheat, Pepper, Day, Power, Light, Anger. Astronomy, Kshatriya, Evolution, Copper Glory (Prakasham), Forest and Mountainous Region, Hamam (Religious and Sacrificial Fire), Temple and places of Worship, Bilious, nature, Bones, Dark-Red or pink Tiger, Deer, Ruddy Goose (Chakravakam), Ayana (six months), Sattva Guna, Grishma Ritu (Jyeshta-Ashada), 50 years, Trees inwardly strong and tall and mountainous trees, Wool, Male, Coarse cloth, Pungent taste, Dhatu (mineral), Square shape, Upward look, Saurastra,

Kalinga, Padyami and Ekadashi Avarga (Alphabets Aa to am), 72 miles, Numerals 1 and 4, Ivory, Fuel, Tall grass, Manikyam (Ruby), Kashyapa Gothram, Old aged, Scents (Gandha), Medium height, Sanskrit and Telugu, Hot Food and Drinks, Tejo Tatwa, Right Nostril (Nava Dwara), Owns Simha, 0 to 20 degrees of Simha is Moolatrikona, Exalted in Mesha (highest point 10 degrees), Marks on Hip.

(B) MOON

Mother, Female Deity or Shakti, Ganapathy, Healthy meals, Cloth, House, Delicate Constitution, Cerebral disorder, Consumption, Asthma, Lungs disease, Watery diseases as Cholera etc., Phlegmatic Disorder, Left eye, Umbrella and Fan, Service under King (rather Queen), Pearls, Corals Bell metal, Butter, Rice, Salt, Fish, Washer man, Sandal and Flowers, Chamaran Fruits, Bathing, Epicurean tastes, Textile fabrics, Marine products, White, Vaisya, Northwest, Mind, Tenderness, Agriculture, Gems, Cows, Women, Bodily happiness, Beauty (Rupa), Blood, Wind, Flem, Parvathi, Hare, Antelope, Crane, Deer, Greek Partridge (chakra), Muhurthem (48 minutes), Sattva Guna, Varsha Ritu (Shravana-Bhadrapada), Taste (Panchendriyas), Marks on head, 70 years, Creepers, Sappy and Blossoming trees, Jala Tatwa, Herbs, Marshy Places, New Cloths, Jewel, Moola (Vegetable), Dwitiya and Dwadashi Yavarga (Alphabets Ya to Ha), 2 miles, Numerals 2 and 7, Curds, Ghee, Milk, White gingerly, Honey, Hotels, Thread business, Retail shop, Dropsy Jaundice, Sugar-Cane, Lotus, Numerology, Literature, Round shape, Plantain tree, Spleen Roots and Bulbs, Sattva Guna, Atreya Gothram, Childhood, Sambrani (a Sandal),

Dwarf, Tamil, Cold Food and Drink, Left Nostril (Navadwara]), Night, strong Face, Owns Kataka, Moola Trikona degrees in Vrishbha are from 4 to 30, Exalted in Vrishbha (Highest point 3 degrees).

(C) KUJA [Mars]

Younger Brothers and Sisters (After Coborns), Lands and Houses, God Subramanian, Anger, Prowess, Courage, Bravery, Brute Force, Energy and Emotion, Fire-Cooking and Engines, Wounds and the fire accidents, War, Red color [Blood Red], Tur Dhal, Engineering and Mechanical Skill, Commandeering, Immoral practices, carbuncle, South, Groundnut, small pox, Glutton, Mensuration, Surveying History and Geography, Hanuman, Earthy profits, Kindred [Gnyathi], Small weapons and instruments, Thief, Enemy, Falsehood, Bilious, Young, Marrow of the Bone and Flesh, Gold-smith, Ram, Coca, Jackal, Monkey, Vulture, One Day, Kshatriya, Thamo Guna, Grishma Ritu [Jyeshta Ashada], Sight Nava Dwara, Marks on back, 16 years, Thorny trees, Male, Sama Veda, Singed cloths, Bitter taste, Dhatu/mineral Square shape, Avanti, Shashti, Kavarga (Alphabets ka to Gnya) Numeral 9 & 63 miles, Lime Burning of Brick and Tiles, Kerosene Oil, Match box, Rugs, Guilting and polishing Police Building work, Chemistry, Logic, Small trees, Bhardwaj Gothram, Youth, Triangular shape, Dwarf, Telugu, Hot Food and Drink, Tejo Tatwa, Mouth (Navadwara), Night, strong Chest, Owns Mesha and Vrischika 0 to 12 degrees of Mesha is Moolatrikona, Exalted in Makara (Highest point 28 degrees).

(D) BUDHA (Mercury)

Maternal, Uncles, Education and knowledge, Vaishnavism and Dualism (Dwaita Philosophy), Vishu Trade, Pure Mathematics and Accounts, Messenger and postal service, Charioteering, Wit and Humor, Grammar, Intelligence, Writing Work, Architecture, Secular knowledge, Firm expressions, Emeralds, Green-gram, Plantains, North, Telugu, Ethics, Speech, Fine-Arts, Upasana, Oyster shell, Place of Recreation, Relatives, Heir-Apparent (Yuvaraj), Friends Sisters, children, Green, Vatha Pithaa Capha mixed (Three diseases), Skin, Artisan, Garuda, Chaathaka Parrot, Cat, Ritu (2 months), Vysya or Trading class, Rajas Guna (passionate), Prithwi Tatwa, Sharad Ritu (Aswija - Kartika), Adopted son, Smelling (Panchendriyas), Marks on Armpit, 20 years, Fruitless Trees, Astrology, Eunuch, Atharvana Veda, Playgrounds, Wet cloth, Mixed Diet [Shad-rasa-priya] Circular shape, Magadha, Saptami Tavarga (Alphabets Ta to Nna) Numeral 5 & 12 miles, Typewriting, Printing, Sculpture, Painting, Drawing and Brushwork, Children, Roots, Leaves, Vegetables, Spleen Meemamsa, Brass, Atreya Gothram, Childhood, Arrow shape, Camphor, Tall, Medium temperature Food & Drink, Left ear (Nava Dwara), Strong both Day and Night, Owns Mithun and Kanya, 16 to 20 degrees of Kanya is Moola Trikona, Exalted in Kanya (Highest point is 15 degrees).

(E) GURU (Jupiter)

Issues, Absolute Brahman, Spiritual knowledge, Intuition and abstract thinking, Yoga practice, Concentration and Meditation, Divine knowledge, Spiritual leadership, Ashta maha Siddhi, Preaching,

Intellectual Involution, Intelligence, enquiry and experiment, Governing and ministerial management, Royal patronage, Titles and paraphernalia and honors, Sruti(Veda), Smrithi (Traditional Law), Scripture, Ethical and moral codes, Philosophical Researches, Patience, Devas and Brahmins, Conquering of the Senses (Jithendriya), Unifying process from diversity to unity, Psychism, Genius and Mental Prodigy, Sanskrit, Eternal truth and Justice, Self-Reliance, Originality, Town Life, North-East, Roots and plantains, Gold and Sapphire, Cats-eye, Sweet Bengal Gram, Sacred Rivers and places of worship, Treasury, Money lending and Banking, Yellow, Fat, Aswath Tree (Pipal Tree), Astrology Preceptor / Sanyasi, Chief, Pigeon, Horse, Swan, One month, Sattva Guna, Akash Tatwa, Hemant Ritu (Margshira-Pushya), Hearing (Panchendriyas), Marks on shoulder, 30 years, Fruit bearing trees, male, Rigveda, Medium, Cloth, Circular shape, Sindhu, Tritiya and Ashtami and Thrayodashi Thavarga (Alphabets from Tha to Na) Numeral 3 & 81 miles, Sambashive (Both Iswara and Parvathi), Cocoanut Tree, Angirasa Gothram, Brain, Youth, Tall, Cool Drink and Food, Right Ear (Nava Dwara), Day-strong Stomach. Owns Dhanush and Meena, 0 to 10 degrees of Dhanush is Moolatrikona, Exalted in Kataka [Highest point 5 degrees).

(F) SUKRA (VENUS)

Wife and conjugal life, Fame and Titles, Temporal glory and Sensual enjoyments, Females and Sexual enjoyment, even during day time, Erotic and Female worship, Company of Prostitutes, Epicureanism, Self-gratification and Lust Music, Dancing and Drama (Triple

Symphony), Sandal, Flower and Aromatics Musk, Civet, Cots, Beddings, Curtains, and all such Paraphernalia, Company of Princesses and Bondir, enjoyments, Beauty, youth, Sensual, Lustrous, Beautiful and Amorous eyes, Wealth and Splendor, Equipages and Cars Garlands, Flowers and Bouquets, Flags and Honors, Flashy Style and Princely living, Law, Literature, Poetry and Dramatic works, Diamonds, Rubies and Silver Vessels, Sea Trade, Navy and Marine, Occupations, Exotic products (foreign), Foreign Ideas and Fashions, Optimism and Survival of the fittest, Sree Vidya Upasana, Desire to rule over mankind, Friendship, Cows, Milk, Curd, Beans, Tamarind, Cupidity and Apollo worship, Mammon (Riches), Deductions and Experimentalism, Charming, Speech, Minister, Marriage and other auspicious Celebrations, Virility, South- East, Peacock, Parrot, Paksha (15 days) Brahmin, Rajo Guna (Passionate), Vasanta Ritu (Chaitra Vaisakha), Taste (Panchendriyas), Marks on face, 7 years, Creepers, Sappy and blossoming trees, Silk, Yajurveda, Harems, Excellent cloths, Pearls, Sour taste Dhatu (mineral), Lakshmi, Cow gram (Avare) Keekata Chowthi, Navami and Chaturdashi Chavarga (Alphabets from Cha to Ingya) Numeral 6 & 144 miles, Doctor's profession, Tin and Led, Politics, Octagon shape, Sex organs, Nose , Bhargava Gothram, Middle age, Medium height, White Sanskrit and Telugu, Cool, Food and Drinks Jala Tatwa Owns Vrishbha and Thula 0 to 5 degrees of Thula is Moolatrikona, Exalted in Meena (highest point 27 degrees).

(G) SANI (Saturn)

Life and Longevity Nocturnal habits and ways Satan, Ayanar and other minor Deities, Contrivance and means of employment, A vocation, Profession, Labor, Agriculture Servitude, Buffaloed, Iron, Blue Stone, and Gems Theft and causing wrongful loss, Heavy work and fasting, Imprisonment, Dismissals, Punishment, Garrulousness and shamelessness causing injury Indebtedness, Funeral obsequies or ceremonies, Alcoholic Drinks and Narcotic Drugs Sexual enjoyment with sickly or aged females or low.

(H) RAHU Female, Paiteesha Gothram, Owns, Kumbha, Exalted in Vrischika

(I) KETU (Dragon's Tail)

Sophistry (fallacious reasoning) and false knowledge, Grand Mother (Mother's mother), Cheating and humbugging, Professions, Onanism, (Self Pollution) and Sodomy, Panchama Pallah and Chakra, Classes, Low class habits, Sinful habits and profession, Tannery, Kilns, Butchery, Alcoholic drinks, Emigration and Slavery in foreign lands, Ulcers, Carbuncles, Cancer, Dropsy, Leprosy, Consumption, Dyspepsia, Prosecution, Imprisonment and undergoing severe penalties and punishments, Torchlight, Dacoit, Murder and Infanticide, Jealousy, Hatred, Penury, Self-immolation, (offer in sacrifice as Sahagamma etc.) Idiocy, Hysteria, War, Slaughter, Havoc and death in the battlefield, Epidemics of a very virulent nature, Snakes, Adder (Viper) and venomous reptiles, Treachery and betrayal, Lust, Dark Red Color, Gun powder and pyrotechnics, Touch (panch indriyan), Out caste, Always smoking, Horse Gram, Emancipation, (Vikarti – Vairagya), Gnana Salvation, (Moksha), Brahman, Ganapathy, Upasana, Vydoora, Elephant trade, Watch mechanism, Goat, Kush grass, (Dharba), Left eye, Navadwara, Powerful day and night, Legs, Foreign Language, Average temperature, Food and drinks, Jaimini Gothram, Thamo Guna, Eunuch, Owns Mesha, Exalted in Vrischika

3 STARS

Stars play a very important part in shaping the effects. So their nature and characteristics have to be properly understood. I now deal with them.

1. Sex of Stars

a. Ashwini, Punarvasu, Pushyami, Hasta, Anuradha, Shravana, Poorvabhadra, and Uttarabhadra are Males.

b. Bharani, Kritika, Rohini, Aridra, Ashlesha, Makha, Pubba, Uttara, Poorvashada, Uttarashada, Dhanishta, and Revati are Females.

c. Mrigsira, Moola, Shatabhisha are Eunuchs.

Notes: These are helpful in determining the sex of Issues. For example in my Table of Charts No. 1 Kuja (Lord of 5) is in Lagna Rasi (Dhanush.) Thus Kuja and the Rasi in which he is are males. Kuja is Retrograde. So he should have given more but he has given only two sons and six daughters. Why? Kuja is in Poorvashada star a female star. Hence the result. Thus if you neglect any one cannon the entire reading goes wrong. In fact during Kuja Bhukti he got a female issue. To determine the result the planet, the Rasi and the star will all have to be taken. Among the three that which goes most powerful will predominate. We know how to work the strength of a planet. The strength of the Rasi is that of its lord. The strength of the star is that of its lord (as per Udu Disha lordship).

In the above case the three are Kuja, Guru and Sukra. As per Shadbala Sukra is most powerful.

2. Gandhantha Stars

Starts that commence with a sign with their first pada or end with a sign with their last pada are termed as Gandhantha Nakshatra. They are Ashwini 1, Makha and Moola 1, and Ashlesha 4, Jyeshta 4 and Revati 4.

3. Sristi, Sthithi & Laya Stars / creative – existive and destructive

Dividing 27 stars from Ashwini into groups of threes the first is Sristi (creative) the second is Sthithi (Executive) and the third is Laya or Samhara Star. (Destructive.)

4. Durithamas Star

The Third and fourth Padas of Samhara Star is known as Durithamsha Star which is said to be very bad.

5. Abhukta Moola Star.

The last one Ghati (24 minutes) of Jyeshta Star, the first two Ghati (48 minutes) of Moola are termed Abhukta Moola. Children born then should be abandoned by the parents for 8 years. After 8 years Father may look at him after doing prescribed Shanti and Homam.

6. Adhomukha Stars (That look downwards)

Bharani, Kritika, Ashlesha, Makha, Pubba, Vishakha, Moola, Poorvashada and Poorvabhadra look downwards.

7. Vainashika / destructive Star

22nd Star from birth Star is Vainashika star (Destructive Star) which should be avoided at Muhurthem.

8. Tatwa of Stars as per Prasna Marga

5 Stars from Ashwini belong to Prithivi Tatwa.

6 Stars from Aridra belong to Jala Tatwa.

6 Stars from Uttara belong to Tejo Tatwa.

5 Stars from Jyeshta belong to Vayu Tatwa.

5 Stars from Dhanishta belong to Akash Tatwa.

Notes: Planets situated in the stars of their own Tatwa will give their individual effects without modifications. But if they are in inimical Tatwa they give untoward effects. For example, If Budha as lord of 5th Bhava be in any of stars of Akash Tatwa his begetting the issues is as distant as the Earth is from Akash. If Sukra as lord of 7 be in any of the stars of Tejo Tatwa his conjugal happiness is as inimical as Water and Fire enemies. Here Sukra is Jala Tatwa Planet.

9. Stellar Geographical Countries

Stars in triplicates from Kritika represent Central Eastern, South eastern etc.in order.

Kritika, Rohini, Mrigsira- Central Provinces- (Panchala)

Aridra to Pushya - Eastern Provinces (Magadha)

Ashlesha to Pubba- South - East Provinces (Kalinga).

Uttara to Chitta Southern Provinces (Avanti)

Swati to Anuradha South west Provinces (Anartha)

Jyeshta to Poorvashada- Western Provinces (Sindh.)

Uttarashada to Dhanishta - North- West Provinces (sowveera.)

Shatabhisha to Uttarabhadra- Northern Provinces (Howra).

Revati to Bharani - North East (Madra and Kowninda.)

Notes: Planets indicates both at birth time and in transit the Geographical places of happening of the effects during their periods. Malefic transiting these stars except in the case of planets who are also re-presentative of those countries cause harm to those countries while Benefices do good always. As stated before the stronger of the two at birth i.e., the planet or the Starry -Lord, gives out the effect.

10. Stellar Castes

Kritika, Pubba, Poorvashada, Poorvabhadra are Brahmins.

Pushya, Uttara, Uttarashada, Uttarabhadra are Kshatriyas.

Ashwini, Punarvasu, Hasta, Abhijit, are Vaisyas.

Rohini, Makha, Anuradha, Revati are Shudras.

Jyeshta, Mrigsira, Chitta, Dhanishta are servicing class.

Aridra, Swati, Shatabhisha, Moola are Butchers.

Bharani, Ashlesha, Vishakha, Shravana are Chandala.

11. Stellar Nature:

Rohini, Utara, Uttarashada, Uttarabhadra are fixed Starts.

Savana, Dhanishta, Shatabhisha, Hasta, Swati are Movable Stars.

Aridra, Ashlesha, Jyeshta, Moola, are (Thishna, Heaty Stars.)

Bharani, Pubba, Poorvashada, Makha, Poorvabhadra are ferocious Stars

Ashwini, Pushya, Hasta are (Laghu) light Stars.

Mrigsira, Chitta, Anuradha, Revati are Soft Stars.

Kritika, Vishakha are Soft and Heaty Starts.

Notes: During the period of a planet in fixed stars he remains at one place while in the movable star he will be always moving. Similarly his character (Soft or ferocious) must be determined.

12. Stellar Parts of Body.

Kritika	Head	Anuradha	Stomach
Rohini	Forehead	Jyeshta	Right side of Trunk
Mrigsira	Eye Brows		
Aridra	Eyes	Moola	Left Side of Trunk
Punarvasu	Nose		
Pushyami	Face	Poorvashada	Back Side of Trunk

Ashlesha	Ears		
Makha	Lips & Chin	Uttarashada	Waist
Pubba	Right hand	Shravana	Sex organs
Uttara	Left hand	Dhanishta	Anus
Hasta	hand Fingers	Shatabhisha	Right Thigh
Chitra	Neck	Poorvabhadra	Left Thigh
Swati	Chest	Uttarabhadra	Shins
Vishakha	Breasts	Revati	Ankles
		Ashwini	Upper parts of Foot.
		Bharani	Bottom part of Foot.

Notes: These are useful in predicating the spread of diseases from part to part of being located in only one part from start to finish. For this you must consider Navamsha chart. in chart 1 of the table of illustrated charts Sukra lord of 6 (house of disease) being in Rohini 1 gives his Amsa to Ashwini 2 which in turn gets back to Rohini 1 again. Thus his disease spread between Rohini (Forehead) and Ashwini (Upper part Foot). First the disease started on the face and finally settled on the upper part of foot. In cases of Vargottama starts the disease remains at the same part always. In case of other starts moving forward six times all the concerned parts are affected ion succession and finally settles at the part indicated by the last star. See my theory of Mathematical Navamsha Diagram.

Now I give the consolidated lists of stellar characteristics.

(1) Ashwini

Ashwini Devas, Male, Prithwi Tatwa, Vaisya, North- East Provinces, Bottom of Foot, Rajali Bird [Garuda]

(2) Bharani

Yama Deva, Female, Prithwi Tatwa, Viswamitra Gothram, Chandala, Crow, North - East Provinces, Bottom of Foot.

(3) Kritika

Agni Deva, Female, Pea-cock, Prithwi Tatwa, Brahmin, Central Provinces, Those who live by fires such as potters, Smiths etc. Army commanders, Skilled Magician, and Meta physician, Diggers, Barbers, Sacrificial Rites Priests, Astronomers Head.

(4) Rohini

Brahma Deva, Shudra, Female, Central Provinces, Devout men, Merchants, Rulers, Rich men, Yogis, Drivers, Men possessed of Cows,

Cattle and Watery animals, Farmers ,Wealth derived from mountain produce, Prithwi Tatwa, Agastya Gothram, Forehead.

(5) Mrigsira

Soma Deva, Eunuch, Serving Class, Central Provinces, Quadrupeds, Somayajees, Yaga, Performer, Revered men, Perfumes, Dress, Pearls, Flowers, Fruits, Precious stones, wild beasts, Birds and Dear, Singer, Lascivious, Good writers or painters, Prithwi Tatwa, Eyebrows.

(6) Aridra

Rudra Deva, Female, Butcher caste, Eastern Provinces, Oil mongers, Washer man, Thieves, Watery products, Delighted in killing and torturing, lying Cheating and tale bearing Thieving, Adultery, Black magic, Sorcery (witchcraft), Exorcism (expelling evil spirits by ceremonies), Pod grains (seed covered), eyes Jala Tatwa.

(7) Punarvasu

Adithi Deva, Male, Vaisya, Eastern provinces, Truthfulness, Generosity, Cleanliness ,Respectable, Decent personal beauty, Sense, Fame, Wealth, Merchants dealing in excellent articles, Fond of service Company of painters and sculptors, Jala Tatwa, Nose.

(8) Pushyami

Brihaspati Deva, Kshatriya male, Eastern Provinces, Singing and Dancing parties ,Bell ringers, Criers (of parka), Yavana, Tradesman

Deceitful men, Forest, Barely Dealers, Wheat, Rice, Sugar, Cane, Forest Produces, Ministers or Rulers, Living by Water, Sadhus Delighted in Sacrificial Rites, Pupil tree (Aswath), Jala Tatwa, Face.

(9) Ashlesha

Sarpa Deva, female, Chandala caste, south East provinces, Water creatures, Serpents, Reptiles, Poison, Medicine of all sorts, Cheating others of their property, Pod grains, Perfumes, Roots, fruits, Palmyra Tree, Vaishistana Gothram, Jala Tatwa, Ears.

10. Makha

Pithru Deva, Shudra, East Provinces, Noted for filial duty, Decedents of Vasista, Acting up to Vedic principles, Elephants, Horses, Religious Rites, Store houses, Merchants, Grains, Wealth, Valiant, Female hater, non-Vegetarian, Female, Aala Tree, Jala Tatwa, Lips and chin.

11. Pubba

Aryama Deva, Brahmin, South East Provinces, Juice Sellers, Prostitutes, Virgins, Dance, Music, Painting, Sculpture, Trade, Will be forever in the enjoyment of the vigor of youth, Cotton, Salt, Honey, Oil, Female, Jack Fruit Tree, Jala Tatwa, Agastya Gothram, Right hand.

12. Uttara

Bhaga Deva, Kshatriya, Southern Provinces, Chaste Women, Bow Makers, Dancers, Ascetics, Kings, Mild, Cleanliness, Modest, Heretic, Generous, Learned, Grains, Wealthy Virtuous, Company of Princes, Jaggery, Salt, Water, Female, Atti Tree, Tejo Tatwa, Agastya Gothram, Left hand.

13. Hastha

Savita Deva, Vaisya, Southern provinces, painters, Well diggers, Barbers, Hillman ,Thieves, physicians, Weavers, Elephant keepers, prostitutes, Garland makers, charioteers, Chief Minister, Merchants, pod grains, Learned in Sastras, Bright appearance, Birds, Male, Atti tree, Tejo Tatwa, Agastya Gothram, Fingers of Hand.

14. Chitra

Trista Deva, Serving class, Southern provinces, Painters, Writers, Singers, Prostitutes, Men learned in Vedas, Dealer in Gold, Women, Various utensils, Jewels, precious stones ,Fine cloths, Manufacturer of perfumes, Mathematician, Wavers, Surgeons, Artists, Oculists, Female, Bilva tree, Tejo Tatwa, Agni Gothram, Neck.

15. Swati

Vayu Deva, Butcher caste, South West Provinces, Servants, Merchants, Boatmen, Reporters, Messengers, Charioteers, Mariners, Dancers and the like, Weak, friendship, Abstentious habits (temperate

), skilled. Tradesman, Birds, Deer, Hoarse, Grain, Beans, Female, Tejo Tatwa, Indra Gothram, Chest.

16. Vishakha

Indra Agni Deva, Chandala caste, South West Provinces, Cotton, Gingelly, Beans, Saffron, Shell - Lac Crops, Everything of bright red or crimson color, Trees yielding red flowers, and red fruits, Black gram, Chick peas, Female, Tejo Tatwa, Breast.

17. Anuradha

Mithra Deva, Sudra Caste, South West provinces, Ministers, Drivers, Bell Ringers ,Friends, Valiant, Head of Parties, Fond of company of Sadhus, Vehicles, Every species of crop, Male, Tejo Tatwa, Stomach.

18. Jyeshta

Indra Deva, Serving class, Western provinces, Chief Ruler, Kings, Chaplain Kings- Favorites, Valiant soldiers, Mixed crowd of men, different castes, Beautiful persons of Good Decent, wealthy and famous, Disposed to cheat others of their property, Fond of travel, Crops, Rain, Female, Vayu Tatwa, Vaishya Gothram, Jack fruit tree, Right side.

19. Moola

Niruti Deva, Butcher caste, Western Provinces, Druggist, Medicinal plants and herbs, Heads of men, Soldier, Rich, Garden work, Fruits,

Flowers, Roots, Seeds, Eunuch, Vayu Tatwa, Pulasthya, Gothram Left Side.

20. Poorvashada

Aapa Deva, Brahmin, Western provinces, Well Wet fields, Rivers, Dealers in Roots and Fruits, Fruits and flowers of water, Creatures of water, Boatmen, Sea Voyage, Earth work, Wealthy and cleanly, Truthful, Gentle manners Female, Vayu Tatwa, Pulasthya Gothram, Banni Tree, Back.

21. Uttarashada

Viswe Deva, Kshatriya, North West Provinces, Diseases, Chief Minister, Wrestlers, Elephants and Horses, Soldiers, Religious Men of Principle, Happy Bright appearance, Female, Vayu Tatwa, Vasista Gothram, Jack fruit Tree, Waist.

22. Shravana

Vishnu Deva, Chandala Caste, North West Provinces, Ears and hearing of News, Public officials, Chief Brahmin, Priests, Physicians, Cunning, Active habits, Efficient Workman, Bold, Virtuous, God fearing, Truthful male, Vayu Tatwa, Vasista Gothram, Kelada Mara Genitals.

23. Dhanishta

Asta Vasu, Serving Class, North West provinces, Heretics Treasury Officer, Shameless, Weak Friendship, Women haters, Generous, Rich, Free from temptation, Female, Pulasthya Gothram, Banni Tree, Akash Tatwa, Anus.

24. Shatabhisha

Varuna Deva, Butcher Caste, Northern provinces, Drunkards or Dealers in liquors, Physicians, Poets, Tradesman, Ministers, Fisherman, Fish and hogs, Washer man Wine, Birds, Eunuch, Aakash, Tatwa, Agastya Gothram, Right thigh.

25. Poorvabhadra

Ajaikapatha Deva, Brahmin, Northern provinces, Thieves, Shepherds, Tortures, Wicked mean Deceitful, Virtue less, Neglecting, Religious Rites, Successful in fight, Male, Cocoanut, Tree, Akash Tatwa, Pulasthya Gothram, Left Thigh.

26. Uttarabhadra

Ahirbudnya Deva, Kshatriya, Northern provinces, Fruit and Roots, Dancers, Travelers, Woman, Gold, Sacrificial rites, Generous Devout Rich Observant of the rules of Holy orders Heretics Ruler, Dealers in Rice, Male, Akash Tatwa, Pulasthya Gothram, Shins.

27. Revati

Pausha Deva, Sudra, North East provinces, Travelers, Servants of reigning, Sovereign Crops of Sharath Ritu, Barbarous, Water flowers, Salt Gems, Conch shells, Boatman, Female, Ippe tree, Akash Tatwa, ankles.

Nakshatra Gandas (Afflicting Stars)

It may be definitely understood that mere starry effects will not totally harm the persons concerned. For candid judgment the merits of the horoscope have to be scrutinized. Yet I find that many still cling to the old conceptions which have almost become family sayings than Astrological Dictums. For those who have still a strong belief in such sayings I narrate below the effects of birth under some Starts with a warning the people may not run amuck at any bad effects portrayed below as they are not always gospel truths. It is after all one of the many secondary tests meant for those that have no mastery of the Astrological Science.

Thithi and Nakshatra Gandas are supposed to affect the parents while Lagna Gandha may affect the native expect when Guru is in Lagna or aspects it. I call Gandam as Bad hereinafter.

For Persons Born in:

The Effects are:

1. Ashwini First Pada: bad to father for 3 months Shanti Gifting gold.

2. Bharani Third Pada: bad to child for 27 days.

3. Rohini First Pada- bad to maternal uncle:

Rohini Second Pada- bad to father:

Rohini Third Pada- bad to Mother:

Rohini Fourth Pada- Good for all:

4. Pushyami:

i.) Male Birth in Pushyami star and Kataka Lagna during day time: bad to father.

ii) Female birth in Pushyami Star and Kataka Lagna during night time- Bad to Mother:

Pushyami First Part--Bad to Maternal Uncle.

Pushyami second part middle-Bad to Maternal Uncle for 3 months.

Shanti-Gift of Sandalwood.

5. Ashlesha First Pada   -   Auspicious.

Ashlesha Second Pada   -   Bad to Child

Ashlesha Third Pada   -   Bad to Mother.

Ashlesha Fourth Pada   -   Bad to Father.

Shanti for all the above Padas-Charity of food. Some opine that Ashlesha star harms Mother- In-law.

6· Two Ghati (48 minutes) duration at Ashlesha end, Makha beginning, Jyeshta end and Moola beginning Revati end and Ashwini beginning-are said to be bad Nakshatra (Gandhantha Stars).

Elders seem to opine that Shanti to Navagraha (Nine Planets) has to be done if a Child is born at these intervals

Notes:-The above Stars are none but those that end with a sign or commence with a sign.

7. Makha First Part: Bad to Father for 5 months
         Shanti is Gift of Horse·

         Rest of the Three Padas are auspicious.

8. Uttara First and Fourth Padas are bad to parents and Coborns for 3 months·

Shanti is Gift of Gingelly in a Vessel.

9. Chitra first, second and third Padas are bad to parents and Coborns for 6 months.

Shanti is Gift of Cloths.

9 (a) Girl born in Vishakha fourth part afflict her husband's Coborns.

10. Jyeshta Star: Divide its span into ten equal parts·

First part is bad to Maternal Grand Mother.

Second part is bad to Maternal Grand Mother

Third part is bad to parents and those Coborns of Mother.

Fourth part is bad to Coborns.

Fifth part is bad to Native.

Sixth part is auspicious to all.

Seventh part is bad to Wife.

Eighth part is bad to Native.

Ninth part is bad to Father.

Tenth part is bad to Mother

Jyeshta Fourth Pada is bad to Father for 9 months.

Shanti: Gift of Cow.

Some opine that a girl born Jyeshta Star hurts her husband's elder brother.

11. Interval of half a Ghati (12 minutes) from the last portion of Ghati (6 minutes) of Jyeshta Star to the first portion of Ghati of Moola Star is known as Antharala. If a male be born in this Antharala it is bad to the child. But if a female be born there is no dosha. (Bad.)

12. Moola Star: Dividing the span into 12 equal parts, if the birth be in the following parts respective relatives and matter suffer:

1) Father 2) Mother 3) Brother 4) Sister 5) Father in law 6) Coborns of father 7) Coborns of Mother 8) Financial Loss 9) Loss of living 10) Poverty and bad to servants, (11 and 12) Native suffers.

These bad effects rule for 3 months.

Shanti is Gift   of   He - buffaloes.

Moola First pada is bad to Father.

Moola Second pada is bad to Mother.

Moola Third Pada is loss of money.

Moola Fourth pada is good for all.

13. Poorvashada:

1) Son born in Day time in Poorvashada Star and Dhanur Lagna causes bad to father.

2) Son born at Sunrise, Sunset or Midnight in:

(a) Poorvashada Star and Dhanur Lagna or:

(b) Pushyami Star and Kataka Lagna will experience bad to himself.

3) One born in Poorvashada Star and Dhanur Lagna, Pushyami Star and Kataka Lagna causes bad to father.

14. Revati Fourth Pada: bad to father for 3 months. Santi Gift of Gold.

CHAPTER III   HOW TO CAST A CHART, AYANAMASHA, EPHEMERES, PANCHANGA, LAGNA, DATE OF BIRTH, TIME OF BIRTH, PLACE OF BIRTH, TATWA THEORY, BHAVA, CUSP

In this chapter I deal with the preliminaries to be observed in the casting of Birth chart and its complimentary charts. Some of the astrologers proceed straightaway on the details furnished by their consultants and some start giving predictions on the Rasi chart furnished by them without even verifying the correctness of the positions of the planets and Lagna even as per the time of Birth furnished by them. This is a first blunder. The most important first set of factors that determine the correct casting are the Date, Time and Place of Birth. Next follows the Ephemeras and Table of Houses and lastly the Ayanamsha to be followed. First I take up the question of Ayanamsha.

1· AYANAMSHA

A lot of diversities of opinion exist on this issue. But these can be very easily solved not by jugglery of interpretations basing on hypothetical presumption but by research and practical application. Very many reputed Astronomers and Scholars have been trying to fix up the correct Ayanamsha but still no two agree whole-heartedly. For, they are arguing only on Astronomical basis and not by verification on the Astrological side. A conference of learned Scholars held at the Government of India Level have decided the issue finally and this has proved to be perfectly true in all cases I have so far handled. That establishes the correctness of this Ayanamsha. As per this the Ayanamsha on 01-01-62 is 23 degrees 19 min. 23 Sec.

No useful purposes will be served by further controversy.

Over this issue settled correctly and intelligently our thanks to those learned scholars. At least I would not differ. Anybody deviating from it is sure to blunder. That is my final say about it. For, it is not my peremptory opinion that I am forcing on my friends but is the opinion of those scholars put to practical test. It is my earnest appeal that my friends may at least hereinafter (if they have not done so, so far) follow this Ayanamsha.

2. EPHEMERES OR PANCHANGA

Next comes the question of Ephemeres. In India there are several Ephemeres and Panchangas. Very many of them are not dependable for accuracy· I have seen astrologers referring to any Panchanga, available at their places. For horoscopy only Drigganitha to be resorted to. See Shloka 4 Adhyaya XIX of Mantreswara's Phaladeepika. Among Panchangas : there are several categories-Drigganitha, Siddhantha (even in Siddhantha there are Several modes) and Vakya, One's faith in his own Panchangam is so deep-rooted that it was sometimes difficult for me to convince them, Anybody following a system other than Drigganitha is sure to falter at the very outset. By Experience I First recommend Raphael's Ephemeres and Table of Houses taking the said Government of India measure of Ayanamsha

Lahiri Ephemeras may also be referred as Narayana Positions are readily available. Tamilians may refer to Kumbakonam mutt Viswanathan Srowthi Panchangam. In olden days when readymade tables were not in existence an Astrologer had to study for several years the subject of Astronomy and then become an Astrologer. He had to cast the positions of planets from the most elementary principal. Thus to complete on horoscope it would take several days, The present age is far advanced in that the mathematical portion is very much eased so much so in a couple of minutes you can cast a chart to minute points. Allan Leo has published Almanac up to 2000 A.D. in advance with weekly positions. There are ready made Tables by Mr. V.B. Ketakar of Bagalkot facilitating the correct positions of planets for any moment by original calculations and this may take after all couple of minutes. All these are said only to impress on the Readers that one need not be Mahamhopadhya - doctorate in Astronomy for success in Astrology.

(3) Lagna (Ascendant)

In the calculations of Lagna there are different schools. Some take the Equatorial spans for all places. This is wrong. Other take Latitude fixed and symmetric spans. This is also not correct. A few take the Latitude varying spans and this is the correct method. This is called Naveen Sputum (New Spans). Even with this you may not be highly accurate. You will please note that even a minute's difference in Lagna or Planetary position is likely to capsize the entire reading. The best and safest method would be to calculate reckoning the sidereal time of Birth with the aid of Table of Houses.

This is so far as mathematical side is concerned. Even this may not be the actual position. For the birth time itself may need correction. As per my theory I go by the other way: the reverse way of establishing the Birth Time after fixing up the exact positions of Lagna by verification of past events.

(4) Birth Time

Controversy as to whether Jalodyam (Puncture of Placenta), Shirodaya (Appearance of Head) or Bhoopathana (Severance from mother's womb) has to be considered may be disposed of by preferring Bhoopathana in Kaliyuga. There are other factors that hinder from knowing the exact moment of Birth, The lady in chamber may not announce the correct time. The exact Sunrise, latitudinal and longitudinal differences of Lagna etc. may all tend to drag down the real point. The responsibility now rests with the Astrologer to fix up the correct time and not try to tell the consultant, if his readings go wrong, that the birth time is wrong. Thus rectification of Birth time is of paramount importance. I have perused many theories on this subject. But none gives proper and all time satisfaction. At last I have been able to find a way out. "Tatwa Siddhantha" and verification of past events from "Phal Kundali" (Division Charts) are the only proper Yardsticks to measures it. The former may give a wider range but the latter will fix it up to the last minute surpassing even the result arrived by Mathematical process.

(5) Tatwa Theory or Theory of five Elements

In the course of universal creation. Brahma (the creative God.) follows a particular principle. Men & Women, Animals and Plants are created by him at specified moments in a day depending on Tatwa of the moment. Hence it is necessary to know more about Tatwa Theory. There are five Tatwa called "Pancha Tatwa" that move in a particular order of prescribed durations, the first cycle commencing with the sunrise on that day. They move first in clockwise order and then in anti-clockwise order the former known as Aroha Tatwa and the latter half cycle being Avaroha Tatwa. Thus one complete cycle comprises first Aroha (Ascending) and then Avaroha (Descending) half cycles

Tatwa in Aroha order	First Tatwa on week day	Duration in minutes	Sex
Prithwi (Earth)	Wednesday	6	Male
Appu (Water)	Monday & Friday	12	Female
Tejas (Fire)	Sunday & Tuesday	18	Male
Vayu (Air	Saturday	24	Female
Akash (Ether)	Thursday	30	Male

. The order of the Tatwa in Aroha series in Prithwi (Earth), Appu (Water), Tejas /Agni (Fire), Vayu (Air) and Akash (Energy) and in

Avaroha series are Akash, Vayu, Tejas, Appu and Prithwi, 6,12,18,24 and 30 minutes are the durations of the Tatwa from Prithwi onwards (in Ascending order) respectively:

Thus in a day of 24 hours there are 8 complete cycles comprising of 8 half cycles of Aroha series and 8 half cycles of Avaroha series. As stated above the starting point of cycle is from Sunrise of the place on that day, the nature of the Tatwa Starting with Sunrise being dependent on the week day as mentioned above in a table.

Note: Hindu and Astrological weekday always commences from Sunrise. The above Table is self-explanatory. Still to make it clearer, I cite an illustration. On Sunday or Tuesday the first Tatwa at Sunrise is Tejas / Agni / Fire ruling for 18 min. Next is Vayu / air lasting for 24 minutes then follow Akash / Energy 30 min. Prithwi / earth 6 min. and Appu / water / Jal 12 min. Thus completing one half cycle of Aroha Tatwa lasting 90 Minutes. Next in series Avaroha Tatwa starts in the order of water 12 minute, Earth 6 minute, Energy 30 Minutes, Air 24 Minutes, and Fire 18 Minutes. Thus completing the second half of first cycle. Then as before Aroha Fire Tatwa shall start and continues in the above said order.

In Support of this theory Shlokas from "Jataka Phala Chintamani is quoted as below:

Analambvagni Bhoovyoma Jala Vayvadhipaha. Khagaha

Kramatharkadayo Vare Swaswakala Pravarthakaha   //

Bhoomyadi Pada Ghatika Vriddhisyadardha Yamake

Yamottarardhe Thadhrasatharohakshavarohanam Rohanam   //

Parivrittidwayam Yame Prathi Praharameedrisam   //

Sthree janma jala Vayvosyadbhoonabhognishu. Pum Janihi   //

Ethena ghatika Gyanam thena lagnam Vinir. disheth   //

As per this shloka, you find that there is Aroha and Avaroha cycles. All the other points of this shloka are the same as described above.

In Aroha Tejas/ Fire-   Vayu/Air-   Akash/ Energy-   Prithwi/ Earth-   Appu / Water is for 18 Minutes, 24 Minutes, 30 minutes, 6 Minutes and 12 Minutes respectively.

Avaroha means Water 12 Minutes, Earth 6 Minutes, Energy 30 Minutes, Air 24 Minutes and Fire 18 Minutes. You see it is just reverse order.

For the given Birth time work out Tatwa and see if the sex of that Tatwa is the same as that of the native. Otherwise make slight-adjustments to fit in properly Male is born in Male Tatwa and female in Female Tatwa. Sometimes you may experience some exceptions to this general rule. At the exact culminating point where the course of Tatwa changes from Aroha to Avaroha or from Avaroha to Aroha sometimes sex opposite to that of the culminating Tatwa will be born. Such cases will be generally of mixed character. If that culminating

Tatwa be a male one it is womanish; if it be female Tatwa Malish; will be born.

Experience shows that sometimes female is born in Akash Midpoint and male in Vayu Midpoint.

For example, on Saturday at the end of one and half hours after sunrise Aroha Teja ends. The very next moment there is the culminating point at which Avaroha Teja commences. A birth at this moment may give rise to Malish Female as Teja Tatwa is male. Again at the end of 3 hrs. When there is a turn from Avaroha Vayu to Aroha Vayu a womanish male may be born as Vayu Tatwa is Female. The Tatwa theory runs still further to minute divisions as Tatwa Antara Tatwa and Antharanthara Tatwa. A male is always born when the major minor and sub Tatwa are all males. A female is born when all these three divisions are females. At other period's animals, plants, birds, reptiles and all non-human births take place. The method of working Antara and Antharanthara Tatwa is similar to that of Disha Bhukti calculations. Here one aspect has to be noticed. The span of one cycle of Tatwa i.e. 3 hours should be the Denominators in calculation the Antara Tatwa and not one and half hours as you have to consider both Aroha and Avaroha Tatwa as belonging to one cycle of Disha. If you want to know the Antharas of an Aroha Tatwa proceeds in ascending order and then in Descending order. If you want the Antharas of an Avaroha Tatwa proceed first in descending order and then resume ascending order. I know this is too taxing a calculation. If you feel this is a hard task you may for the present stop with the main Tatwa as you get the rest rectified with the aid of Division charts. Please see chapter on Division Charts.

For example, if the Antharas of Aroha Tejo Tatwa has to be worked out order of Antharas would be Aroha Tejo, Aroha Vayu. Aroha Akash, Aroha Prithwi, Aroha Appu and then Avaroha Appu, Avaroha Prithwi, Avaroha Akash, Avaroha Vayu, Avaroha Tejo. Thus the entire spank of 18 minutes of Tejo Tatwa has to be proportionately distributed among all the ten Tatwa proportionate to their spans.

6. Special Effects of Tatwa:

1) Differences in effect exist between Aroha and Avaroha Tatwa. By the very concept one born in Aroha Tatwa will experience gradual rise in all aspects as ages advance while one born in Avaroha Tatwa may experience declining effects.

2) One born in Prithwi Tatwa is always earthy in his ambition of life and materialist while those born in Tejo Tatwa become very impressive, imposing and powerful personalities. Great Statesmen Politicians, Engineers etc.

3) A combination of Aroha Teja, Simha or Mesha Lagna with powerful and well placed Sun and Kuja backed up by their Disha at proper age bestows very high and brilliant opportunities in life and he will become a Commandeering personality.

4) Akash Tathvites have the faculty of deep thinking, analytical and Researcher mind. Great thinkers, Philosophers, Scientists, Inventors are usually born here. They will not have much attachment to earthy or material matters.

7. How to Cast Chart:

There are two methods of casting a chart one with the aid of Standard Panchanga and the other with a standard Ephemeris containing position of Planets in degree and minutes. For my method I recommend the latter as it gives a more correct position. For the benefit of students of Astrology I quote the two methods.

(a) Panchanga Method

In Panchanga the starry position (Nakshatra pada) of planets are given only when there are changes in Padas. Period of Retrogression (Vakra), Stationary (Sthambana), Eclipse (Astha), Fast movement (Gathi Chara) are also shown. All these conditions of Planets will have to be carefully noted as they upset the normal effects. For details please read Chapter on Shadbala. By noting the Chara (Star Part) of the Planet prior to the time of birth fix the Planets in the Zodiac. To do this you must know the parts of Stars located in a sign. This is the first lesson in Astrology which can be learnt from any elementary book. Still to make this book self-sufficient, as contemplated by me I give the most general rules. Starting with Ashwini fix up 9 Padas (parts) of stars in succession in each Rasi starting from Mesha. Thus Mesha has 4 Padas of Ashwini from 0 to 13degree-20minutes, 4 pada of Bharani from 13-20 to 26-40 and 1 pada of Kritika from 26degree-40minutes to 30 Degrees. Thus you see that span of star is 13 -20 degrees while that of a pada is 3degrees -20minutes. By working you will see that the last star Revati situated from 16-40 to 30 degrees of Meena. On a careful analysis of the way in which the stars are distributed in the signs you find the following noteworthy points which will be helpful for detailed readings.

a) Some stars are wholly situated in a sign (whole stars).

b) Some are spread half and half in two Rasis (equal stars)

c) Some are unequally spread in two Rasis (unequal stars).

d) Some commence a Rasi (Commencing stars)

e) Some terminate in a Rashi (ending stars):

Notes: Each one of the above has its own characteristic.

To find the exact position in degrees etc. of a Planet work out by proportional method for the birth time taking the period of transit of that pada as Denominator.

To find Moon's position work out from the span of the Day's Star (Birth Star) for the Birth time by proportional method, similarly you get the Thithi, Yoga and Karana for the birth time.

To fix up Lagna work as follows:

At the end of the line of the day's details the expired or in some Panchangas the remaining portion of the Lagna at Sunrise (sign occupied by Sun) is given in Ghati. Referring to the table of Rasi - mana (spans) for the birth latitude calculate the balance of that Rasimana at sunrise. Add to this spans of succeeding signs till you reach the birth time. You get the Lagna. By proportions you can work out its exact position in degrees considering the Rasimana of that Lagna Rasi being equal to 30 digress.

Note: Planetary positions for the all latitudes and longitudes at a particular moment may for our purposes be taken to be same while Lagna fixing depends on both the Latitude and Longitude.

Till now I have said of Panchanga method now I will describe Ephemeres method.

B. Ephemeras method

This is easier as it gives more details as Sidereal time, Declinations (Kranthi), Latitude and other readymade tables of great help. In the Ephemeres the positions are given for each day for a particular time. Raphael gives Sayana positions at noon G.M.T. which is 5-30 P.M. I.S.T. Lahiri gives Narayana positions at 5-30 a.m. from 1941 onwards and at 5.30 P.M. I.S.T. before that year. Work out by proportions the exact positions of all planets.

To work out Lagna

Add to the sidereal time (at noon) of the day the birth time in L.M.T. measure. If birth be before Noon (I.S.T.) deduct 12 hours... The net is the sidereal hour of Birth. There are other minute rectifications prescribed here. But you may not worry with all that as the difference will be only in seconds which does not affect the Lagna point perceptibly as also with all those corrections you may not, for several reasons stated already, arrive at correct Lagna point which has to be finally settled by other methods that I narrate later in my chapter on Division Charts. So, for our present purpose this will do. With the aid of Table of Houses for the latitude of the birth place find out the Lagna point and tenth cusp. Then the seventh and fourth bhava cusps will be 180 degrees apart from Lagna and tenth cusps respectively. Thus you have struck the four major Radix points the cusps of 1, 4, 7 and 10 Bhavas. In fixing up the other cusps we have to differ from the western theory and adopt the Hindu system enunciated in Sripathi Paddathi. Divide the span of each quarter into 3 equal parts and thus fix up the other Cuspal points. As calculations under Sripathi Paddathi is hard nut to crack for all beginners and also takes a lot of time to work out a horoscope those working with Panchanga only may after working out the Lagna point find out it equivalent Sidereal birth hour referring to Table of houses and then work out the Tenth cusp for the said sidereal Hours. This can be done at glance.

8. Bhava Kundali (Bhava Chart)

You have now understood to fix up Bhava cusps. The next thing is to establish the Bhava Chart. To do this you must know the total span or spread of a Bhava. Call the middle point between the cusps of the Bhavas in questions and of its preceding and succeeding Bhavas as A and B. Then the spread of this Bhava is from A to B. A is the starting point the cusp its center and B the terminating point of that Bhava. Sometimes this span may be located in one, two or three Rasis.

Illustrations:

1. A Bhava ranging from 1 Degree, of Mithun to 29. Deg. of Mithun (1 Rasi)

2. A Bhava Starting at 20 Deg. of Vrishbha ending at 24 Deg. of Mithun (2 Rasis)

3. A Bhava from 29 Deg. of Mesha to 1 Deg. of Mithun (Case of 3 Rasis)

Having thus fixed the commencing, middle and ending points of a Bhava it is now easy to locate the Bhava positions of planets. A planet located anywhere within the said span falls in that Bhava.

Here you have to note the significance of cusp of a Bhava and its span. The two are used for two different purposes. When we want to find the lordships we have to consider Cuspal point. The lord of the Rasi containing this point is that Bhava Lord. The Bhava positions of planets are to be judged from the Span. To have clear perspective prepare two charts as follows:

Call the first as Cuspal chart and the second as Bhava Chart. In both the Charts retain Lagna in the same sign as in Rasi Chart. In Cuspal chart place the cusps in the concerned signs. Sometimes a Rasi may contain even three cusps and some other time a Rasi may go without even one cusp. Thus a planet may own even 3 or more Bhavas or not own any Bhava.

Next in Bhava Chart place the planet in the Rasi as distant from Lagna Rasi equal to the number of the Bhava in which it is situated. In this chart reckon only the no. of Rasis from Lagna Rasi but forget that they are Rasis.

For Example, if a planet is found to be in fourth Bhava, place it in fourth Rasi from Lagna Rasi. This gives a ready view of Bhava positions.

Next is to erect other correlative Maps - i.e. from General (Rasi Chart) to species (Division charts) which will be discussed in the Chapter on Divisions Charts.

Mere Rasi chart is not sufficient for predication. The world "Rasi" in Sanskrit means a 'Heap' A Rasi (Bhava) heap of many connotations. All of them may not be simultaneously good, bad or mixed. Some of them may go bad, some good and some mixed. How to discern and differentiate them is a knotty problem. By god's grace I have found a way out- so simple and sure that even a layman can grasp easily. Please read the chapter on Division charts.

After Rasi chart the Bhava chart should also be erected. Judgement from Rasi chart only of the Lordships of planets or their Bhava positions merely counting from Lagna Rasi will not always be correct. For real perspective cases where the Rasi and Bhava Charts are similar the Astrologer may be successful even with Rasi prediction. Then that Astrologer must be really running good time. When his bad period rules charts of differences come to him only to pull down his name.

Sometimes a planet appearing to own a Bhava in Rasi chart may not own it. Sometimes some other planet may own that Bhava. One may own even three or more Bhavas or none. Planet appearing to be in Bhava as per Rasi chart may not be actually there where the Bhava Span is scrutinized. In Gochara reading this is of much significance. In Rasi a planet may appear to be transiting a Bhava at a particulars time while as per Bhava span he may be in any of the abutting Bhavas one less or one more. That makes a lot of difference.

To make this more lucid I cite the following chart. I have handled.

Birth on 20-01-1931 at 1.19p.m. Sialkot, Punjab, Pakistan (latitude 33degree N long 75 degree E)

Rah. 25–36			R. Guru 20–47
		RASI CHART	R. Kuja 16–37
Moon 23–3 Sun 6–25			
Sani 23–6 Budha 13 56	Sukra 20–20		Ket

2015/12/17 16:04

XII 27 9	15	I 3 15 II 27	III 21 9
XI 21 3		CUSPAL CHART	IV 15 3
X 15 3			V 21 3
IX 21 9	VII 3 15 VIII 27	15	VI 27 9

2015/12/17 16:04

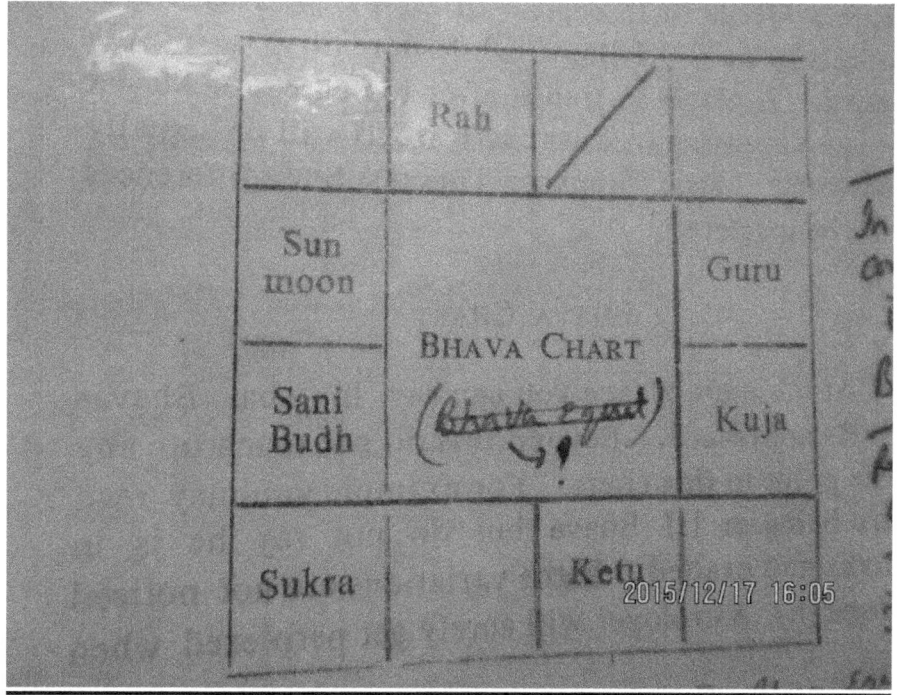

Rashi Chart Lagna is in Vrishbha. In Mithun retrograde Guru 20 degree 47 minute. In Kataka is retrograde Kuja 16 degree 37 Minute. In Kanya is Ketu. In Vrischika it is Sukra 20 degree 20 Minute. In Dhanush it is Sani 23 Degree 6 Minute, & Budha 13 degree 56 minute. In Makara Moon 23 degree 3 minute and Sun 6 degree 25 Minute. In Meena Rahu with 25 degree 56 minutes.

Cuspal Chart See the difference between Rasi chart and this chart, II Bhava falls in Vrishbha instead of Mithun as in Rasi Chart. III falls in Mithun, IV in Kataka and likewise there are several displacements. For example if you read Budha as lord of 2 and 5 form Rasi chart it goes wrong as really he becomes that lord of 3 and 6 Bhavas (see Cuspal chart) Again 1st and 2nd Bhavas fall in Vrishbha. No Bhavas exist in Mesha and Thula. Thus you see a lot of difference here. Predictions made with such Rasi charts are sure to go wrong.

Bhava Spans **Those** marked in small figures are the end points of Bhavas. For example XII Bhavas spreads from degrees Meena to 15 degrees of Mesha. In Rasi Rahu appears to be in XI Bhava while in reality is in XII Bhava. A planet in transit after 15 degrees of Mesha though appearing in Rasi Chart in XII will be actually transiting Lagna Bhava. These subtle differences must be mastered.

BHAVA CHART At the mere glance you can say in what Bhavas Planets are placed. Beyond this do not touch up any other point in this chart. For example you may read Guru being in III Bhava but do not say he is in Kataka and exalted. If the variations are not noticed in time the Astrologer will surely get perplexed when he sees effects caused by unconcerned planet who do not appear to have any jurisdiction in the Rasi Chart. Such disappointments are likely to mislead us to resort to some other Rule or even coin a new Rule or resort to a different system of Disha. That is how this science is actually butchered. Plenty of such novel sayings may be seen "Uttarakalamrita;' supposed to have been written by Kalidasa evidently and definitely not that famous Kavi Kalidas as may be seen from the poetic composition and Motely Miscellaneous matter of various Dictums being compiled. Also no authority is forthcoming on the point of Kavi Kalidasa to have written any book on Astrology.

9. Finally I deal with Disha Calculations As explained before I follow Udu Disha. (Vimshottari of 120 years) The following table gives the details:

Stars	Disha (Lord)	Disha (Years)
Ashwini - Makha - Moola	Ketu	7
Bharani - Pubba - P. Ashada	Sukra	20
Kritika - Uttara- U. Ashada	Sun	6
Rohini- Hasta - Shravana	Moon	10
Mrigsira - Chitra - Dhanishta	Kuja	7
Aridra- Swati - Shatabhisha	Rahu	18
Punarvasu- Vishakha- P. Bhadra	Guru	16
Pushyami Anuradha- U. Bhadra	Sani	19
Ashlesha - Jyeshta - Revati	Budha	17

Notes: The order of the Disha are the same as tabled above. The balance of Disha at tabled above. The balance of Disha at birth is got by Rule of three process in proportion to the balance of span of Birth Star.

To calculate Bhukti. Multiply the Disha year by Bhukti Luard's years. Leave off the unit digit in the product. The rest of the figure represents months. Thrice the unit digit represent days. This is a shorter contrivance of the mathematical process of distributing the Disha years to all the 9 planets in proportion to the Disha years of the respective Bhukti Lord. By continuance of this process you may work out Antara, Antharanthara, Sookshma, Prana etc. to minutest parts, for our purposes stage up to Antara is enough.

For example say Guru Disha Sani Bhukti: -   Sani years are 19 and Guru Years 16. Multiplying both. 16x19. we get 304 leaving off the unit digit 4 the remaining figures is 30 months. Multiplying the unit digit 4 by 3 we get 12 (days.) So Guru Disha - Sani Bhukti or for the same reason Sani Disha - Guru Bhukti is 30 months and 12 days or 2 years, 8 months, 12 days. If unit digit to Zero then no days.

To make the subject more clearly I illustrate by an example.

Birth on 01-02-1962 at 4.00 P.M. Or Ghati 2323 Thedi 19, AT Madras.

Sun rise is 6.39 A.M.

First work out from Panchanga

(Shorwti's)

The Starry quarters prior of Birth time are:

Sun	Moon	Kuja	Budha	Guru
Sravan-3	Moola- 1	U-ashad-3	Dhani -2	Dhan-1

Sukra	Shani	Rahu	Ketu
Sravan -4	Saravan-1	Ashlesha -3	Dhan-1

Budha and Shani Are Asta (Eclipsed)

Moon :   On 01.02.62 is Jyeshta 21-15 Gh (03-09 p.m.)

on 02.02.62 is Moola 18-14 Gh (01.57 P.M.)

The span of Moola star is roughly 57 Ghatis. So each Quarter is 14 and ¼ Ghatis.

Birth being at Gh 23-23, time passed in Moola Star till birth time (23-23) - (21-15): Gh 2-8. This is just the first quarter of Moola. So birth star or the Chara of Moon is Moola 1.

To find the Disha Balance

For 57 Ghati the birth Disha of Ketu is 7 years.

For GH 2-8   (2and 2/15) it is   7/57X17/5 yrs. i.e. 0-3-4.

So the balance of Ketu Disha at birth is yrs. 6-8-26.

As per this Moon's position is 30 minutes in Dhanush (8-0-30)

TO FIND LAGNA

On 01-02-62 the balance of Udaya Lagna (Rising Sign Makara) is given as Gh. 1-46. As per Rasi span of Madras Latitude Balance in Makara is 1-46, Kumbha 4-16, Meena 4-10, Mesha- 4-28, Vrishbha 5-3 total Gh. 19-43. Birth being at 23-23 Gh. Then there still remains Gh. 3-40 to pass in Mithun. Thus Lagna is Mithun.

To work out the exact Degree Position of Lagna

The span of Mithun is Gh. 5-29 or 329 Vig. for 30 degrees. for Gh. 3-40 or 220 vig. the position is   (30X200)/329 = Degree 20-03 Hence the exact Lagna point in Mithun Deg. 20.3 (2-20-3)

To work out tenth Meridian from Lagna

On pp. 72, of Lahiri Ephemeres of 1962 in the Table of Houses: - against 20-3 of Mithun in Col. 1, the sidereal time under Madras (col 4) is 0-35-50.

Under col. 2 (10th House for all places) this Sidereal time of 0-35-50 is seen across 18 degrees of Meena in col. 1. Thus the cusp of the Tenth House is 18 degrees of Meena.

Ephemeres Method

Now let us get them verified by Lahiri Ephemeres 1962.

At 5-30 a.m. on 1-2-60 the position of:

Sun is   9-18-19

Moon is   7-24-28

Day's motion of Sun is 61. Min and that of Moon Deg. 14-18. From 5-30 a.m. to 4 p.m. (birth time) the interval is   10 and ½ Hours

(Working by Rule of Three) their movements of 10   and ½ hours. Sun advances by 26 min. and Moon by Deg. 6-15. Adding these their actual positions at birth time are 9-18-45 and 8-0-43 respectively. For this position of Moon the balance of Ketu Disha at birth is yrs. 6-7-15. The other planetary positions at birth time may be similarly worked out and they are:

Kuja	R. Budha	Guru	Sukra
9-6-19	9-27-18	9-24-27	9-20-3
Sani	Rahu	Ketu	
9-10-4	3-26-39	9-26-39	

Budha is Retrograde and Set (eclipsed). Sukra is set. Guru is set, Sani is set. Take always Mean Rahu position on pp 30 and not True Rahu.

In this connection I have corresponded with Mr. Lahiri to include daily positions of Mean Rahu.

If you understand working with Logarithmic Tables on pages 74-75 and Disha working on pages 68-69 of this Ephemeres you can simplify the Arithmetical workings.

Please remember that the three figures used to name the Zodiacal position of planets etc., are Rasi - Degree- minute. If it be measure of period years months days but of time Hours minutes seconds.

From the above degree position you may evaluate their starry positions. In this case the two method viz. Panchanga and Ephemeres agree as the two under reference are following the same method and same Ayanamsha (of Govt. of India) with all that will show you slight difference between the two When that is the case how can we rely on all Panchangas or Ephemeres.

Calculations of Lagna from Ephemeres and Table of Houses:

On 1-2-62 Sidereal time at noon. is Hrs. 20-43-58

Add birth time   4-0-0

Total   24-43-58

Less correction for local time of Madras. 0-9-0

Net   24-34-58

Say roughly Hrs.   24-35

When it exceeds 24 hours subtract 24 hours similarly, when birth is before Noon subtract 12 hours. Now we get 0-35 from the Narayana Table of Houses on pp, 72 under col. Madras corresponding to Sd. time of 0-35 in col. 1. Lagna is shown as 20 degrees of Mithun. Again under col. 2. Tenth House 18 degrees is seen against 0-35. So Tenth House cusp is 18 degrees of Meena.

10. Essentials to be noted in Horoscopic Epigraphy

I have seen samples of horoscopic writings where in long list of description of the days are given but very many of those details are not at all made use of. Sometimes what is most essential (say birth place) is left out. That is not proper, The Epigraph should contain all essential elements that are used. Other unnecessary items may be left out. I now narrate what are those essentials.

1) Date of Birth:

In the case of Birth from Sunrise to Midnight English data and weekday gives correct clue, but in the case of birth from Midnight to Sunrise confusions may arise if definite indications are not given. For the date and the weekday of the English calendar are from midnight to midnight while the Hindu or Astrological weekday is from Sunrise to Sunrise. To be more specific it is better to state such night births as between say from example. Thursday the 1st February the 2nd February 62. or say so many Ghati or hours after Sunrise on Thursday

1-2-62. Strictly speaking such a birth say at 1 am. Should be recorded as on 1 a.m. on Thursday day the 2-2-62.

2) Time of Birth:

Better it stated in Indian standard time I.S.T. which is 5   and half hours in advance of Greenwitch mean Noon time (G.M.T.). If I.S.T. is not followed in any place specify the details to correlate with any known measures (G.M.T. preferably). In India from 1-9-1942 to 15-10-1945 the I.S.T. was advanced by one hours for war purposes. Recordings of Birth during this interval should give details of old & advanced measures. In all cases the safest is to state in interval from Sunrise stating the time of Sunrise you have reckoned. For, in some cases such recordings may not be quite correct. On page 4 of Lahiri Ephemeres you find some - more details on observance of timings.

3. Place of Birth

This is of utmost importance for striking sunrise & Lagna. If you can give the Latitude and Longitude of the place it would be enough. Otherwise name the nearest important Town to gauge them. To find the longitudal difference in the time multiply the degrees of Longitude by 4. You get in terms of minutes. That gives the difference in time from 5.30 (I.S.T.). For example Bangalore is 77   degree 34   minute. This multiplied by 4 gives 310 min. or 5 hours 10 min. The standard time in India corresponding to G.M.T. Noon being 5.30 p.m. the corresponding L.M.T. at Bangalore is got by deducting 20 min.

Similarly in cases of Longitudes higher than 82   and ½ degrees yielding times greater than 5-30 you must add the difference to get the local time.

4. Sunrise

Note the Sunrise of the place on the day of Birth in I.S.T. If you cannot get it for the birth place you can evaluate by adding or subtracting (as the case may be) the longitude difference in time between the Birth place and a known place of which Sunrise may be struck correctly. As Sunrise and Sunset are more or less the same on the same date of an English month in all years you may refer to any year details for this purpose.

5. Week Day

For Astrological purposes always reckon Week days from Sunrise. For the evaluation of Week day see my Commentary on naming of Week days by the theory of Hora.

6. THITHI   **7**. YOGA   **8**. BIRTH YOGA POINT

9. BIRTH YOGA PLANET   **10**. BIRTH AVAYOGI
11.DUPLICATE YOGI   **12** BIRTH STAR   **13** KARANA
14. ZERO RASIS   **15**. BIRTH TATWA   **16**. LAGNA.

This is the Chief Key for success in predication. For, as may be seen later even a minute's difference in Lagna capsizes the entire reading. So you must fix it up not by strict mathematical calculations by direct method but by a reverse process of Verification of some of the important past events with aid of Division Charts. Please read my chapter on Division Charts. (Vide Part II New Techniques & Predictions. / Astrology Part 2)

17. Tenth Meridian Cusp.

18. Graha Samyam (at least of the operative Disha Lord).

19. Planetary positions in Degrees and Minutes with nothings of their special conditions such as Retrograde, Eclipse, Stationery, Fast movement, etc.

20. The star quarter positions (Nakshatra pada of planet and Lagna)

21. Any one Disha Chart with Starters and Rulers.

22. All the Division Charts.

23. A rough estimate of progressed Moon on any day.

For Correct predications the above details are essential. With a little practice you will be able to evaluate these factors say in about an hour.

I find Ephemeres of Lahiri more convenient to use.

Now a days computer soft wares are available to fix all astronomical calculations. I find Parashara software better than others. But still I have seen it is not accurate as far as cusp or D-11, D-5, D-16 Charts. So still I have to follow Lahiri Ephemeres. In most cases like profession or marital life, or wealth charts etc. this software takes care including Birth yogi and Ava yogi and Main Disha periods and sub periods.

CAPTER IV

Chapter IV   Shadbala (Six fold Strength of Planets)

In this chapter I deal with only salient features of Shadbala that are just necessary and sufficient for the estimation of quantum of effects. Be it remembered that things said in this chapter do never speak of the nature of effect (good or bad) but merely augment the degree of effect. Even here there are two schools of thought. One saying that a planet endowed with strength does good while the other school merely supports the point of quantum of effect. By research I agree with the latter class. My method of deal on this subject is again straight, and simple. I do not enter deep into the intricacies. Those that desire to include more on this subject may study Sripathi Paddathi and tell me in the end where they stand.

Before venturing with details you must know that difference between the words "**Shadbala**" and "**Shadvaraga bala**". The former is the strength countable in any Varga (Varga Means Division) while the latter refers to some total of strengths in all Vargas which is rather a very cumbersome working not of much help in arriving at candid conclusions. As stated above Sripathi Paddathi gives a lot of mathematical calculations to find out Saptha - Vargajabala and Ista Kasta of Planets which is tough, cumbersome and hard nut to crack. If you work out a horoscope as per this I am afraid you cannot handle many horoscopes in life time. Above all such consolidated of planets in terms of their strengths only to read such effects as from comparison of their strengths. For example the Shaka of Brahmin may be found out from the strongest planet. When two or more planets combine to find the strongest one and in such other readings based on comparative study. But for individual effects and singled effects this will not be helpful. To makeup this deficiency I have introduced Division Charts which are handy, intelligible easy to grasp and have revealed wonderful truth appreciated by all. Hence I confine myself to Shadbala.

Shadbala or Six fold strengths are: 1) Kalaja or Temporal 2) Chesta or Motional. 3) Ucchaja or Exaltation, 4) Dik of directional 5) Ayana or Declination and lastly 6) Sthana or positional. Let us take up one by one.

1) Kalaja or Temporal: They are of three kinds:

a) Ahoratra Bala (Night & Day): Moon, Kuja and Sukra are strong at nights, Sun, Guru and Sani are strong at day time. Budha is strong at all times.

b) Paksha Bala: Benefices are strong is Shukla Paksha (Bright-Half)

Malefics are strong in Krishna Paksha. (Dark -Half)

c) Planets who become the lord of the year, Month, Day and Hour (Hora) get ¼, ½, ¾, and 1Rupa strengths respectively. Here I have mentioned the figures not to make use of them direct but only to show the comparative strengths,

Explanations: -

1) The Lord of the year is the Lord of the year (for Word) of Birth. This may be found from the Panchanga of the year. This is no other than the Lord of the weekday on which the Lunar year of birth commenced.

2. The lord of the month is the Lord of the sign position by Sun at Birth.

3. Lord of the day is Lord of the Birth week day.

4. Lord of the hour is the Birth Hora Lord.

From the above it is evident that among all the rest Hour Lord becomes the most powerful. It is so because that is the minutest division among them.

2. Chesta Bala (Motional Strength): Moon gets Chesta bala in Shukla Paksha (Bright Half) Sun in Uttarayana (Northern Couse)

Others when Retrograde.

3. Uccha Bala (Exaltation): -

Highest strength at highest point and Zero at its opposite point. Intermediary positional strength has to be calculated by proportions.

4. Digbala (Directional):

Sun and Kuja are strong in Tenth, Moon and Sukra in Fourth, Budha and Guru in Lagna, Sani in Seventh.

At opposite Rasis they lose their entire strength (p.s.) here only Rasi positions have to be considered and not the Bhavas.

5. Ayana Bala (Declinations):

Sun, Moon Guru and Sukra have Ayanabala in Northern Declination.

Moon, Budha and Sani have Ayanabala in Southern Declination.

In planetary war that of the Northern Declination becomes victorious.

Sripathi Paddathi says Budha has Ayanabala always.

Notes: a) A planet gives out its effect during it's:

i) Disha Bhukti

ii) The year for which he becomes lord.

iii) The Ayana represented by his Declination at Birth.

iv) His month (when sun transits signs owned by him)

v) His Paksha (Just as you find Paksha for Moon. Similarly find for all planets)

vi) His week Day.

vii) His Thithi (See planetary Karakatwas)

viii) His Hora.

ix) His Lagna.

x) His star.

b) Readers may kindly note that different types of results arise from the above six types of strength. While giving out prediction this point should be remembered. For example during the period of a planet having:

Kalabala: One will have success in matters where time is of importance.

Chesta bala: In matters where motion or movements are concerned.

Uccha bala: For all matters.

Dik bala: One with Digbala will always command others.

Sthanbala: For position, Status etc.

Using these factors in the Rasi Chart alone may not give you full satisfaction as read form Division Charts. As stated already they do not indicate the nature (Good or Bad) of effects but merely measures the degree of effects good or bad to be ascertained by other tests.

CHAPTER V - YOGADHYAYA

In this chapter I deal with the Yogas cited in standard tests. Please note that the mere satisfaction of the tests of yoga in a horoscope is not sufficient unless backed by proper Dishas at proper period in one's life. If the Disha Periods of the planets causing these Yogas do not at all intervene he may not at all enjoy their effects. It will be like dreaming Kingship. Suppose those dishas come in his childhood or teens or in old age. Even then it does not help him much. Thus the period of intervention of the planetary Disha has special significance. The existence of a yoga no doubt may be seen to be operative throughout life, but its magnitude is very little. Its exuberance is seen only during its Disha. Exception may be in the case of Nabhasadi Yogas and Malika Yoga. With this short preamble I will get into the subject proper.

1. Ruchaka, Bhadra, Hamsa, Malava and Sasayogas are said to be formed by Kuja, Budha, Guru, Sukra and Sani respectively, occupying in a Kendra identical with its own or exaltation house. Then the effects of each one of them is narrated. There is no need to remember all these effects as many a time these readings do not fit in properly. It is after all particularization of the general principle that a power full placed planet in auspicious Bhava confers good results, akin to the natural and functional character of the planet. By functional is meant the Bhavas it owns, its situation and Aspect. It is these effects that are felt during the period of the planet in question. So it is very necessary to master the natural characteristics of planets, Rasis, Bhavas, and Stars etc. For these read chapter II. Coupling these you can give as meticulous a details as is needed. I do not like the way of elaboration. I prefer concise ideas and mathematical and logical analysis. So I always try to strike at fundamental basic theories.

Notes: The author suggests to consider the above Yogas from Moon also like Lagna, I do not relish it for the reason that if Moon be placed in Dushtana from Lagna what purpose would it serve reckoning from such unhappy Moon. Also, this sort of double reckoning from Lagna and Moon increases the possibility of the existence of these Yogas in very many cases -not true to face. Any conclusion drawn from the division is more reliable than broad based ones. Here, between Moon and Lagna the latter is the minutest point. So predictions based on Lagna will be appropriate. So I always prefer Lagna as basis in all cases rather than Moon. I am aware of the saying that whichever of the two is more powerful that should be taken but that stands only in theory but practice shows the other way.

2) Sunapha, Anapha and Duradura Yogas.

When planets other than Sun occupy 2 or 12 or both 2 and 12 houses from Moon the Yogas are respectively called Sunapha, Anapha and Duradura. All sort of readings are given here, but I view differently.

110

These Yogas will not be effective unless those planets are powerfully situated as being in own or exaltation house. I will then read that the portfolios of the concerned planets become favorable during their periods. By Portfolio I mean their natural and functional traits.

Note: Here no differentiation is made between a Benefic and Malefic planet. That means that any of the seven planets may be there.

P.S. In all these Yogas leave off Rahu and Ketu to commence with. After deciding the existence of the yoga read the modified effects due to Rahu and Ketu if they are also there.

3. Vasi, Vesi and Ubhayachari

They are of two kinds Benefic and Malefic. Benefic planets other than Moon occupying 2 or 12 or Both from Sun cause Shubhavesi, Shubhavasi and Subha Ubhayachari, respectively. All these give good effects of course akin to their traits as stated before.

Malefics in the above positions cause Ashubha Vesi, Ashubhavasi and Ashubha Ubhayachari all bad.

4. Karthariyoga

Benefices in 2 and 12 from Lagna cause Shubha Karthari - Good if only 2nd from Lagna is occupied by Benefices it is called Sushubha also good.

Malefics on both sides of Lagna form Papakarthari bad.

5. Sankya Yoga

Note: i) Jaimini narrates extending the above theory of Karthariyoga. Planets on either side of Lagna placed at equal distances (Counted in Rasis) cause the above yoga. If they are Benefices good effects results. If Malefics the results are bad i.e., they must be in 2-12; 3-11; 4-10; 5-9; 6-8; 1-7; Rasis from Lagna.

ii) Their Quantum of effects has to be judged by their Shadbala.

5. Adhiyoga

i) Benefices occupying 6, 7, and 8 houses from Moon cause Adhiyoga. A commander or Head of a Town or one a Superior position is born.

ii) The above reckoning made from Lagna also gives the above effects as it is nothing but Shubha Karthari Yoga of the partner.

iii) If Malefics occupy the said positions bad result emanate.

P.S. In all the above Yogas from 2 to 5 there is a similarity of Rule which may be generalized as follows. Benefices on either side of Sun, Moon, Lagna or the Seventh House, confer good results when Malefics in such position do bad. This is nothing but general theory of a planet or Bhava getting hemmed in between Benefices or Malefics. When narrating a principle all contingencies arising thereon must be solved. A mere ordinary position of these yoga causing planets is different from strongly placed ones. The conjunctions or aspect of any other planet on them will modify the effects.

Nothing is said of the Sun and Moon being in such positions. My Research has related that if they be on either side of Lagna they become protectors and do lot of good.

Even Malefics in own or exalted Rasis in the above positions tend to do good more definitely if they become Birth Yogis.

6. Kesariyoga

Guru in Kendra position from Moon causes this yoga. Special Shadbala positions do more good than mere ordinary position.

Here again a point arises. Of what good is that Kesariyoga when either Moon or Jupiter is badly posited? In fact these anomalies arise in all Yogas caused by reckoning from Sun or Moon. I have seen the effects of Kesariyoga existing in majority of cases. They are in very ordinary positions. For real effect Guru must be strong, well placed and be the owner of auspicious houses and more than all his Disha should operate in time, Even so these Yogas are found to work splendidly in my Division Charts rather than in Rasi Chart.

7. Maha Bhagya Yoga

In the case of a male born in the Day, Sun, Moon and Lagna posited in male signs (odd Rasi) and in the case of a female born in the Night the above three being posited in female signs (even Rasi) Mahabhagya yoga is formed.

This is not sufficient. I add further that they should all be in happy relative positions and their Dishas should intervene. The stronger they are the greater the effect. Sun and Moon are luminaries that play the role of first importance on both Animate and Inanimate objects and Lagna is representative of the native. Day is strong for men and Night for women. If these three factors are happily placed they confer yoga.

8. Vasumatyoga

All the Benefices occupying Upachaya houses (3-6-10-11) from Lagna or Moon cause this yoga. For reasons stated already I prefer Lagna.

9 Amalayoga

Benefices in 10th from Lagna or Moon cause this yoga. This is nothing but particularization of the general theory that benefices in any Bhava do good.

10. Pushkalayoga

If the lord of Lagna in conjunction with the lord of the Rasi occupied by Moon be situated in a Kendra (why Kendra it may be any auspicious Bhava) and aspects Lagna and at the same time there be a strong planet in Lagna Pushkalayoga is formed.

11. Shubhamalayoga

If all the planets (Seven) occupy 5th, 6th and 7th houses from Lagna Shubhamalayoga is formed. The effects are controller of people, extolled by kins, liberal minded, evinces interest in aiding others works, Lover of Relatives, Blessed with good children and wife, Courageous.

12. Ashubhamalayog

If all the planets (seven) are in 6th, 8th and 12th from Lagna - all bad effects.

13. Lakshmi yoga

Lord of 9 (house of wealth) and Sukra (Karaka of wealth) are posited in their own or exalted houses identical with Kendra or Kona this yoga is formed. All good effects of 9th Bhava and of Sukra are felt.

This is nothing but a singled out instance of the general theory of the Bhava lord and the Karaka being happily placed.

I omit the other Yogas as they are all caused by the happy positions of the Luminaries and Planets from Lagna since they come under General Enunciations.

14. Raja yoga

Kendra is ascribed to be a place of Vishnu (giver of happiness and position) while Kona is Lakshmi Sthana (Goddess of wealth). If the lords of these two combine together and are situated in a happy position he will have status, happiness and wealth. Here there is a further discrimination. Of the 4 Kendra 10th is the most powerful one. Of the Kona 9th is more powerful. By their combinations Raja Yoga is formed as 10th is Raja Sthana (House of Government).

15. Mishra yoga

If the lords of good Bhavas combine with the lords of inauspicious Bhavas mixed effects are felt.

16. Viparitha Raja Yoga

If the Lords of Dushtana (3-6-8-12) are in Dushtana Viparitha Raja Yoga is formed. Its effect are Sudden, unexpected and meteoric rise in life a bolt from the blue.

17. Parivarthana yoga (Mutual exchange)

a) Exchange between lords of Kendra's and Kona's good.

b) Exchange between lords of good and bad Bhavas - mixed.

c) Exchange between lords of Bad Bhavas Viparitha Yoga.

18. Kemadrumayoga

I leave off Kemadruma Yoga as its efficacy has not impressed me.

19. Neecha Bhanga Raja Yoga

This is another subject handled differently by different authors. It indicates getting up in position from a lower status and circumstances. Several Dictums are laid down in Texts but still in exhaustive. So I list below the Cannons of judgment as confirmed by practical experience.

By Raja Yoga you should only mean prosperity and good position in life and circumstances akin to that of a highly placed person and should not always be interpreted to mean Kingship or even service in Government. For it may hold good even in the case a Non-Government employee and merchants too.

CANNONS OF JUDGMENT OF NEECHA BHANGA

1. If the lord of the Rasi occupied by the Neecha planet is in Kendra position from Lagna, Moon or itself this Neecha Bhanga is said to be formed. I prefer its position from Lagna only and would even second Trinal positions too. Or

2. If the lord of the Exaltation Sign of this Neecha planet is similarly circumstanced as in 1 above then also this effect. Or

3. If the Neecha Planet is Retrograde the same effect is formed. Or

4. If the Lord of the Rasi occupied by the Neecha planet is otherwise strong by being exalted with Digbala and in good Bhava this good is formed. Or

5. If the lord of the Rasi occupied by the Neecha Planet aspects this Neecha Planet then also this yoga is formed. Or

6. If this Neecha Planet be Negative Planet under the definitions of our New Techniques as explained in part II this good arises.

This Neecha Bhanga effect will be specially seen to start in the Bhukti of the planet causing this Neecha Bhanga under the Disha of this Neecha Planet under consideration. Till then the bad effect of this Neecha planet may be running.

Some opine that this Neecha Bhanga will be caused if the planet who has this Neecha sign as his Exaltation Rasi be in positions as described in 1 above. But I do not second this as it has not proved true in all cases. Logically also it fails to stand true in all cases as in the case of Moon no planet has Vrischika (Moon's Neecha Rasi) as its exaltation sign. Any rule should hold good in all cases without exception and here it fails and so this rule may not be relied upon.

The rest of the Yogas are of ordinary nature. So I leave them off.

I have found that very many horoscopes having some of these Yogas in Rasi chart have deceived the Astrologers and perplexed them. But are these sayings untrue? Certainly not as they are said by Daivaganas. I apply all these to my Division charts where it works out cent-percent correct. Please see chapter on division charts in my part II Edition i.e. Astrology.

Astrology Lessons

CHAPTER VI

FEMALE HOROSCOPY

There is practically not much difference in the handling of male and female horoscopes. What is said of male Horoscope applies equally to female horoscopes too except for the following difference. If she is not an independent earning member living separately then all effects read out of her chart, except a specified few should be ascribed to her benefactor (Poshaka). It is usually said that her husband will have these effects. But I use broader word Benefactor. So long as she remains unmarried (may be for her life time even) she lives with somebody who protects her. During that period of dependency whatever effects (other than touching her person such as health etc.) accrue from her horoscope should be ascribed to her then protector (Ashrayadatha). This is not special to Female Horoscope only. It is generally applicable to all cases male or female so long as they remain as whole time dependents. For example, all the children depend upon their parents. The servant who has been one with the family members depends upon his master. Bear in mind that this is not the case of a paid servant who works for wages and lives by himself. The real significance underlying here is the factor of absolute dependence on whom they entirely rely for their living and upbringing. It may not be always said that the effects of the children will be felt by the parents and that of the wife by her husband. This will be a blind say again. During the particular period (Disha -Bhukti) it is likely that those dependents are supported by persons other than parents or husband. In such cases their horoscopes revelations should be applied to the lives of such benefactors.

With all the generalized commonness between the male and female horoscopes there are a few deviations on some aspects. Her mangalyam (Womanhood as opposed to widowhood) has to be read out from her eight house, issues from the ninth her association Chastity from the fourth and all about her husband from the seventh as usual. Benefics in these places do well while Malefics do harm. Though there be a malefic in the eighth house if a strong benefic planet be in the second house she herself will die before her husband.

It is time old conception attributing a heavy damaging effect on Mangalyam (Husband's Longevity) to Kuja if he be in 1, 2, 4, 7, 8 or 12 houses. This is named Angirasa Dosha. The intrinsic reason of this is that Kuja should not afflict the two important houses seventh (Husband's Welfare) and eight (her mangalyam) either by situation or aspect. To have such an effect Kuja must be in Lagna (aspects both 7 and 8), Second house (aspects 8), fourth house (aspects 7) seventh house (situation), eighth house (situation), and twelfth house (aspects 7). On this analogy I would rather include one more position of fifth house from where Kuja will have special fourth aspect on the eighth house. I do not know why this place is not included. May be for the conception that the fifth house being Kona (Trine) makes him benefic. But his aspect as natural malefic is always bad that too under special aspect, so I prefer to add this position also. Readers will verify this by research.

People may not be led away by the mere positional Character of Kuja. Why particularize Kuja only in fact, any malefic planet may cause harm to any house. What is the specialty with Kuja? You know Kuja is called Mangal meaning he is Mangal Karaka (portfolio planet of Auspiciousness), as mangalyam is one of auspicious nature Kuja is specially pointed out. Even so one will be undeservedly accusing Kuja and thereby bear his course if he fails to know the real merits of Kuja in a horoscope.

For example, for Simha Lagna Kuja in Vrischika (4th house) becomes a first rate Yoga karaka Perchance if he also becomes birth yogi, then where is Angaraka Dosha; rather I read it as Angaraka Yoga. Kuja may be in the above stated position say in the 8th house. Then is he there to effect the partner's life or for any other cause must be dissected. May be there to affect the younger Coborns. His landed property. (Natural Karakatwas) or the connotations of house he owns. Without being able to dissect if one suddenly jumps into ready conclusion it would not only be doing dis-service to the science but bear the curse of not only Kuja for false accusations but also of the disappointed parents of daughters whose marriage alliances are broken by the gospel spell of astrologers on Angaraka Dosha. Even suppose there is real Angaraka Dosha. Should not its Disha operate in time? Suppose it operates after her 70th age how harmful it would be? Or suppose it operate before the age of marriage in teens where is the widowhood for a maiden? Even here I wish to draw a line of difference. Even supposing that there is Kuja Dosha if here mangalyam has to be affected it is only the affliction of the 8th house that has to be considered as the seventh house afflicts only the health, happiness and property of the husband. So the only malefic position of the husband. So the only malefic position for Kuja to affect mangalyam are 1,2,5,8, and 12 only.

One thing I wish to impress on the Reader is that any conclusion should never be made at sight of positions of planets in Rasi Chart. How is that Kuja, what are the houses he owns, is he birth Yogi or Avayogi, is he a positive or negative planet, what is the nature of the Rasi star and Bhava he occupies, is his dosha acting all these must be considered. Many are of the opinion that for union this dosha should exist in both the horoscopes perhaps they are not for the death of only one of the partners if really there be indication of this Dosha in one horoscope it is not the existence of similar Dosha in the other horoscope that should be preferred. In such cases the longevity of the other horoscope has to be investigated. I have been especially lengthy on the point of Angaraka Dosha as I have personally witnessed and also personally experienced much hurdle due to half learned astrologers or purohits having skeleton knowledge of Astrology but in whom some will have placed implicit faith (blind faith) as they are their family purohits. You may prefer to consult an infant Astrologer to a Purohit who may be relied upon only to get the ceremonials performed after the fixture is made.

CHASTITY OF A WOMAN

Fourth house is supposed to indicate the chastity of a woman. But in my experience I have found that other houses also cause effect. Sixth, seventh, eleventh and Twelfth house also are indicators. Generally, Sukra is the chief indicator. Any sort of relationship of Sukra Venus and Kuja Mars is likely to affect the moral standard. Sixth house is the house of disease. If one has to suffer from venereal disease she should be immoral. Seventh being the house of partner and sexual union she may elect an outsider. Eleventh house is the house of increase in Profit and even second wife or husband. A planet of this house may cause here to join one for profits. Lastly, Twelfth house is a house of Bed comforts may be better bad comforts elsewhere. Then, why is fourth house singled out? It is a house of happiness and general conduct. If she feels for increased earthy happiness she may go out of her good conduct or her happiness in life may go down due to her bad conduct. In my experience fourth house has not given me proper satisfactory due as the other houses. May be that in ancient day's sexual immorality being unthinkable even slight bad character in a lady was construed very seriously. These days even high sexual immorality is not of much significance. Then what about the natural bad conduct of fourth house?

While reading the chastity of a woman do not suddenly conclude merely looking at an afflicting planet. See if its Disha or Bhukti operates. For, we have seen ladies remaining chaste till a distant age after which they have changed their conduct. Some who have been immoral in early ages have turned out a new leaf in their later ages. Such malefic planetary Dishas operating before attaining puberty will have practically no effects on them. All such considerations will have to be made before venturing to predict.

Often it is questioned as to who will predecease.

In my experience I have found it difficult to answer with the aid of Astrology. So I prefer Palmistry. For this reading look to the Marriage Line-short horizontal line or lines from the percussion lying on the Mount of Budha between the base of the little finger and Heart-Line. If this Marriage Line bends down towards Heart Line you may readily declare that the partner will die first, else the native only?

Thus whenever I feel that astrology does not give me assured position I prefer palmistry. After all two are sister and allied subjects.

So should female horoscope be scrutinized.

CHAPTER VII

BHAVA PHALAMS RESULTS OF HOUSES

In this chapter I discuss Bhava Phalam in a new style but not transgressing the old Dictums. There are books on this one subject written in great volumes but of no avail. What useful purpose would it serve by editing a book of thousands of planetary combinations? First of all it would be difficult to search in that long series our particular combinations. Secondly, if perchance there be one such it may not give out satisfactory reading. Such publications are mere glamour. I warn the public not to be attracted by such glamour books. I really pity those that commenced such publications. What we really need is the enunciation of definite principles of judgment. If this be known anybody can himself coin even lacs of such combinations-

In the delineation of particular effects of a Bhava or of the time of occurrence of the event etc., a lot of alternative tests are stated· you might have experienced utmost difficulty and bugbear in Prediction if there is even one alternative test. If so, what would be one's fate with a number of alternatives? As far as possible we should try to minimize those alternatives or ambiguities. It is only then that the science becomes really a definite science. To do this you must research and not merely read. Having this in my view I have in fact dealt my chapters of my publication in a lucid way· Where I have experienced doubts and ambiguities or even alternatives I have left them out. Only those that are definite and proved by practical application to be true have been inserted. Now to the subject proper. I take it for granted that the Readers will have gone through the general Rules laid down in texts on this subject. The common principles of judgment of Bhava Phalam must be properly digested to avoid confusions that may arise due to seeming contradictory Dictums. It is said that for the progression or prosperity of a bhava it needs the aspect of its lord. This is just like the master of a house being present at headquarters controlling his family. An aspect should always be judged from Rasi Chart. For strength of aspect you should look to the longitudinal differences between the aspector and the aspectee. Aspects include Conjunction also. All planets aspect the seventh sign and planets in the seventh sign from. In addition there are some special aspects as follows 4 and 8 for Kuja, 5 and 9 for Guru, 3 and 10 for Sani, Rahu and Ketu. Usually we confine to these aspects. It is only under the Tajak, western and in Parasari method of judging the strength of aspects 3,4,5,7,9,10 and 11 Rasi aspects are also taken. These may be considered only for the purposes of estimating Dristibala- strength by aspect and not at other times. I would prefer to call these relatives positions as being auspicious in determination of Disha Bhukti Phalam rather than view on the plane of aspects. This amounts to saying that planets in Trines and Quadrants do good. The second house is a neutral house and a planet there remains neutral in character. He does neither good nor bad. A planet in 3, 6, 8, 12 (Dushtana) is of no avail, except Kuja in 6 as he aspects the Lagna or the Bhava from which it is in 6th, with eighth aspect. All these are said referring both to Lagna, Rasi, and Bhava Rasi in question. Here you witness contradictory sayings, but you will find that they are seeming contradictions and that both the sayings are correct. Instead of general explanation an example will clear this riddle. For a Mesha Lagna birth suppose Moon is in Kumbha. As lord of 4 in 11 house Moon causes all good effects to the native especially for professional income. But the connotations of fourth Bhava may suffer as its lord is in eighth place from that Bhava – his mother may die, his education may suffer etc. Thus there are two angles of judgement, one its position from Lagna causing direct effect to the native and the other to the portfolios of the Bhava when reckoned from the Bhava (Chiefly to those relatives). This is how you should try to reconcile such seeming contradictions.

II.   It is said that whichever Bhava has its lord occupying the eighth place or obscured by solar Rays or in Depression or in inimical house without benefic aspects that Bhava suffers. This is a classic dictum. But in practice all these are not true. But should the saying of Daivaganas or Astrological Servants be discarded? NO. I reconcile it in this way. A bhava connotes multiple portfolios. It may be one, some or all of them going bad. Some of them may still remain good. How to discern which goes good and which bad? They have no doubt suggested to couple up the concerned Karaka planet and then read the effect. Unfortunately even then correct judgements cannot be achieved. Thus I find that the Division Charts is the only way for successful analysis. But one thing is always certain that planets obscured by solar rays (Eclipsed) or defeated in planetary fight (Graha Yudha) can never do good where ever they be and cause worst results when they happen to be in bad positions. There is an exception here. Sani and Sukra give their moiety of effects though eclipsed.

III   **Benefics on either side of** Bhava without the aspect of Malefics promote the bhava while malefic so placed without the aspect of Benefics spoil the Bhava:-

Here a doubt arises whether this theory holds good if those planets are on either side of the Cuspal degree of that Bhava in the same sign or that they should be in the two signs on either side. To commonsense the first sort of situation is sufficient as the principle underlying is protection of the Cuspal point of the Bhava from both sides. But studying the Yogas of Varahamihira it is nothing but an extension of "Karthariyoga" with respect to that Bhava. So it follows that for further effects they should be in different Rasis on either side.

IV. Texts say that a planet in Bhava Sandhi (Junction point) becomes ineffective. This needs proper clarification as otherwise it gives rise to misconception. I say that such a planet becomes ineffective so far as the effects of those two Bhavas are concerned. But he will surely give the effects of the Bhava he owns and aspects and his natural characteristics (Karakatwas).

V. *Measurement of Degree of effects:* A planet at the exact Cuspal degree of a Bhava gives full effects of that Bhava and at the two ends of the Bhava it becomes zero. For intermediary positions work out by proportion. Here you may note one specialty. A planet situated between the commencement of a Bhava and its central point (cusp) is said to be in 'Aroha' (Ascending order) while between the cusp and its end point in 'Avaroha' (Descending order). The two positions though being the same Bhava, give different types of effect. During the period of the Aroha planet you experience progressive effects of the Bhava day by day while in the case of Avaroha its effects go on declining as time advances.

VI. *Planets in Bhavas:* There is diversity of opinion regarding the effects of planets in Bhavas and this gives a good chance for timely interpreters. It is said that Benefics in auspicious Bhavas give good while Malefics do bad and that Malefics in auspicious Bhavas do worse. But regarding Benefics in bad houses (Dushtana) there is diversity of opinion. Whenever one feels doubtful about a point the only way open is to get it settled by practical verification. Unfortunately our people do not take pains for genuine Research work nor have the broad mind to encourage one in the field. In my experience I find that both theories are true. How? Here the angle of judgement should be varied. A planet has two qualities. Natural (Karakatwas) and Functional (Adhipathya). While judging the effects both these traits have to be simultaneously considered.

The crucial existence of Horoscopy is collateral with the existence of philosophy or any branch of exact science except with the difference that the former is based upon intuition and the latter upon experiment. The philosophical principles involved in the subject are permanent and will appeal to the rational judgment of any sober thinker in the world. The ancient Rishis have built the subject upon these basic principles and these will stand as long as any positive science could stand and the one generic term by which we can denominate all these basic principles is called Karakatwas or the connotations of natural truths in planetary symbols. Also the results vary according to the Ascendant of each individual as well as the planetary indications of the good and the bad; and this is denoted in the Adhipathya of the planets. Thus the Karakatwas and Adhipathya represent two poles of thought by which human mind can view the destinies of humanity. These two poles of thought serve as mediums by which we can view the horoscopes:- the one represents the natural and philosophical structure of the horoscope while the other the temporary indications of the weal and woe of mankind. The former course is neglected by all and the latter, if beneficial, is liked by all. Nobody would care for the permanent course of nature but everyone would be very much interested in the ephemeral turns-rather ups and downs of life of indivi-dual prospects with a view to give unqualified satiety to their tickling fancies and greedy appetites. The astrologer should as a rule have these modes of human tendencies in his mind before he begins to predict. The ancient Rishis were great sages and they cared more for spirituality and Aerul entities than for ephemeral existence of the world and the sacrificial and philosophical works they have written for the benefit of mankind are based upon those Aerual principles which would last as long as any science could stand. The means to know and obtain truth are more important than truth itself and verification with our miseries and enjoyments will be still more important.

In horoscopy the permanent means to understand truth have been represented in Karakatwas and the temporary indications of the weal and woe by Adhipathya; and the cyclic times of occurrences of good and bad events are de-noted in the Dishas.

The formation of the horoscope has therefore to be construed in two ways-permanent and worldly. In one word the former is synthesized in planetary Karaka power and the latter in planetary Functions according to the Ascendant. Predictions in connection with the former are not needed by the general public and only the latter one is eagerly sought for, but the Astrologer should be keen enough to draw clear demarcation line between these two sides of Nature and then to make his predictions, and he should be cautious in foretelling events pertaining to both these sides.

To speak specifically, let us take two types of horoscopes. In one horoscope the functional Malefics have become completely powerless by occupying 3, 6, 8 or 12 houses and one or two? Functional Benefics occupy or aspect Lagna or Kendra and Trikona houses. If these functional Malefics chance to be Natural Benefics, the Native will be deprived of the Natural Karaka powers of those planets. Probably these Natural Benefics may indicate good conduct, spirituality, good memory, charity, philanthropy etc. and the Native will have none of those qualities. But all the same these planets give very good yoga results in their periods: and the native in his glory and luxurious environments may not have an opportunity (why not even destined to do so) to understand or even feel the necessity of the above good qualities; nor would he like others saying of them or pointing out these defects. The Astrologer should keep back such matters from the knowledge of the Native as such an action on the part of a professional man is "Off times proof of his wisdom" and unless he is pressed by an earnest inquirer he is not authorized to divulge such secrets-the secrets being the weaknesses of human Nature.

In another horoscope the functional Malefics. (Some of them may be Natural Benefics also) may occupy Lagna, the 5th and 9th houses which may be well occupied or aspected by Natural Benefics also. The Native may further have the disadvantage of bad dishas. The Native may really be pious, god-fearing intelligent, well versed in various sciences and heartily charitable. Yet, the general mass representing rather the lustful generality will not look to these good traits much less eared for them. His sufferings alone will be taken into account, and if he suffers for his past karma he should also be doomed and disliked by the fishy favorites of the world. As said by elders, man is the instrument through which karmic results are given effect to.

The following cannons of judgment are listed for reference.

NATURAL AND FUNCTIONAL TRAITS OF PLANETS

There are four sets of planets as follows:-

1. NATURAL BENEFICS or Benefics by Nature: Those that are good by nature and they are, Guru, Sukra, Budha without the conjunction of natural Malefics and Waxing Moon (Shukla Paksha Chandra). As per our N.T.P. and Western Astrology we classify sun as mostly a Natural Benefic.

2. NATURAL MALEFICS or Malefics by nature   Budha with natural Malefics, waning moon (Krishna Paksha Chandra), Kuja, Sani, Rahul and Ketu.

3. FUNCTIONAL BENEFICS   All planets (whether natural Benefics or Malefics) owning houses (Bhavas) other then 3, 6, 8 and 12 (dustanas)

4. FUNCTIONAL MALEFICS   All planets (Whether natural Benefics or Malefics) owning 3, 6, 8, or 12 bhavas. -

Hereafter we use the following abbreviations for brevity:-

N.B.   Natural Benefics   N.M.   Natural Malefics.

F.B.   Functional Benefics.   F.M.   Functional Malefics.

We classify the effects of these planets as follows:-

a)   N.B. do good to the bhavas they conjoin or aspect while so far as their Karakatwas are concerned they go good while in good bhavas and bad when in Dustanas except in the case of 3 and 6 called Upachayas where gradual good seen,

b)   N.M. spoil the effects of all the bhavas they conjoin or aspect while their Karakatwas prove to be good when in good bhavas and bad when in 8 and 12 bhavas giving gradual good when in 3 and 6 the Upachaya houses.

c)   F.B. Functional Benefics when in good bhavas do good to the bhava it owns while in Dustanas spoil the good of the bhava it owns.

d)   F.M.   Functional Malefics when in good bhavas attain strength to increase the bad effects of the bhava it owns while in Dustanas get weakened to do bad and subsequently cause what is known as 'Viparitha Raja Yoga' meaning sudden and surprise good effects coming quite unexpectedly and without trials. But this attainment may be after hurdles-disappointments and deaths of someone. Please note that in the case of (c) and (d) we have confined to the effects of the bhava it owns and not said of the bhava he conjoins or aspects for which the tests of (a) and (b) only should be applied

How TO READ THE COMBINED EFFECTS OF BOTH NATURAL AND FUNCTIONAL.

This is easy. Read the effects of each separately and narrate both the effects. Sometimes it may seem to give contradictory effects. Those differential effects have to be read out during their periods separately.

e)   Any planet in Exaltation is supposed to be naturally strong and so it's Natural Karakatwas go good. But the good or bad ensuing from Functional character has to be read out as has already been explained F.B. exalted does good to the bhava it owns while F.M. in exaltation harms the bad bhava it owns meaning that it increases the bad effects of that bad bhava.

f)   Whether Benefic or Malefic, Natural or Functional, in own house, does always good to the bhava it owns.

g)   As position in Bhava is more important than the Rasi in which it is, F.B. in good bhava though Neecha etc., retains its good of the bhava it owns - may be of less degree. The same planet though exalted in Dustana will be of no avail.

h)   FB. should not be related with F.M·

i)   Enemies by nature should not be related even though they are F. B. For, their powers to do good will be lessened by counter action.

j)   In the Case of malefic function of 8th bhava exception is made for Sun and Moon. In my experience this may hold good for Maraka effect only (death inflicting) and not for Yoga (material prosperity), Even for Marka other planets may cause death during their bhuktis in the Disha of such a Sun or Moon.

Thus you see that Adhipathya (Functional) is more important than Karakatwas (Natural). Nature and Function should both be mixed up in all cases and results read out by correlating the two.

EXAMPLE: -   Guru is the owner of the eighth house for Vrishbha Lagna. Even supposing that this Natural Benefic occupies Lagna (a powerful and auspicious bhava for him) he will give miserable and disastrous results in his Disha. As a benefic by nature he will give knowledge, piety, name, ordinary earning etc., so far as his Karaka powers are not repugnant to his functional powers.

7. Mesha onwards and Lagna onwards the organs of Kala Purusha and the Native are distributed from head *to* foot. Also the planets are ascribed organs of Body. (See chapter on Karakatwa) Then how to predict the affliction of any part of the body amidst three sets of alternatives. Please note that outward or superficial part of body are connoted by Rasis while inward or deep-rooted diseases are to be read out from Bhavas. In both cases the concerned Karaka planet also to be considered for final confirmation.

EXAMPLE: -   Meena Rasi is kala Purusha's feet. Suppose Rahu is there (or any malefic) then predict that the feet is afflicted outwardly, How to read out the nature of the afflicting planet? If Rahu is there Eczema, or skin diseases· Then look to Rahu who is karaka for feet. If he is there the disease is confirmed on the outer part of leg. Suppose it is also Mesha Lagna then Meena becomes the twelfth Bhava when again the inward part of feet is situated now you can say that both the inward and outward parts of feet are affected.

8.0 Finally I cite an example to illustrate the intricacies of Natural and functional traits

Take Chart No. I of my Illustrated charts (19-6-1907 at 7.9 p.m.)

Here Sani, lord of 2, is in 4. As lord of Finance in Kendra he is good for finance. As lord of 3 (House of short Debts) in 4 he also gave short debt. As natural malefic in 4 he affects general happiness and comforts in life. Sukra as Natural benefic in 6th lessens disease debt and enemies to the native but as kalathrakaraka in 6 caused illness to wife during his period. As lord of 6 in 6 he gave Viparitha Yoga in Sukra Bhukti. But as lord of 11 in 6th caused loss.

(i) DHANATH DHANAM (Lord of 2 in 3):-   With an initial capital increasing one's financial condition.

(ii) BHRATHRUVATH-BRATHRU (Lord of 3 in 5):- Younger coborns by step-mother or her like or Adoptive brother or his like.

(iii) VIDYATH-VIDYA, MATHRUVATH-MATHRU, SUKHATH-SUKHAM:- (Lord of 4 in 7)   Advancing from one type of education to another or from one Degree to a higher Degree. From general to Technical education etc. Having step-mother or her like. Happiness abounding in plenty and in succession.

(iv) PUTRAT-PUTRAM-(Lord of 5 in 9) :-   Adopted children or step children (legal or illegal.)

(v) Lord of 6 in 11   ROGATH-ROGAM, RINATH-RINAM, SHATRU-VATH-SHATRU: -   One disease developing to another. First debt getting increased to further debts. One enemy raising other enemies.

(VI) KALATHRATH-KALATHRAM (Lord of 7 in 1) :- Multiple wives or concubines.

(VII) MARANATH-MARANAM (Lord of 8 in 3):- One death being the cause of a subsequent death as for example Shahgamana or murder by revenge, or Suicide by the death of one lover etc.

(VIII) BHAGYATH BHAGYAM, PITHRUVATH-PITHRU (Lord of 9 in 5):- Starting with original affluence acquiring further afflux of wealth. In the case of the other reading it may be common in western countries where Bigamy for ladies is socially permitted. Among Hindus it is an obnoxious idea-i.e. to have a Bhinna pitru (Secondary father). Yet there are instances of widow marriage or illicit intimacy of the mother with a paramour who will then assume the role of Step-Father especially when he protects him. We are witnessing some cases in this modernized world.

(IX) KARMATH-KARMAM (Lord of 10 in 7) :- From one profession to another. This is self explanatory.

(X) LABHATH-LABHAM (Lord of 11 in 9) multiple profits.

(XI) VYAYATH-VYAYAM (Lord of 12 in 11) Spending and over-Spending.

These effects are to be read only when such a combination is there. You should not try to question conversely as to why such a combination is not there though such effects are felt. In such cases you should try to seek other reasons and not blame this theory.

IX KARAKATH BHAVA

That Rasi removed from a planet at a distance equal to the number of the Bhava counted from Lagna for which it is the karaka of that Bhava-all counted only in Rasi measure is the Karakath Bhava Rasi. This theory is made use of by me only to measure the Quantity of effects of a Bhava confining to Astakavarga. Whenever I say that I confine myself to a particular aspect it does not mean that I am not aware of the other aspects. If I quote a theory it should be applicable to all horoscopes without exception and without alternatives. It is my earnest wish to make this sacred science more mathematical and precise.

Texts say a lot of alternative tests to find out the number of issue, wives, co-borns, etc., but to my utter disappointment all of them have not proved to be generally applicable. So I have finally *preferred* the following tests:

THEORY: Find the number of Bindus (Dots) in the Binnastakavarga of the karaka planet contributed to the

Rasi connoting Karakath Bhava. So much will be the Quantum of effect subject to modifications as follows:-

IF THE PLANET OR ASCENDANT CONTRIBUTING THE BINDU BE:-

a) In own House or own Navamsha-Double the effect.

b) In Retrogression or exaltation in Rasi or Navamsha-Treble effect.

c) In Retrogression and own Rasi or Amsa six times effect.

d) In Retrogression and exaltation in Rasi or Amsa – Nine times.

e) Astha (set) or Neecha (Debilitation) or Zero-Rasi and other sets of combinations have to be guessed by the Readers.

f) When Lagna is aspected by its lord- Double it.

To be more specific and illustrative-to find the number of:

i) YOUNGER CO-BORNS: Consider 3rd house from Kuja in Kujastakavarga.

II) ISSUES: Take the 5th from Guru in his Astakavarga.

III) WIVES OR WOMEN one JOINS : Weigh the 7th from Sukra in his Astakavarga.

IV) ELDER Co-BORN: See to 11th from Guru in Guru Astakavarga.

To impress the genuineness of this theory on the Readers I cite the following example –Refer to chart No. 1 of my series (19-6-1907 at 07.09 p.m.)

YOUNGERCo-BORNS : In the Astakavarga of Kuja (Karaka for younger co-borns) Lagna and Moon contribute Bindus to 3rdhouse (House of younger co-borns) Kuja i.e. to Kumba. Note lord of Lagna (Guru) aspects Lagna. Hence Lagna gives 2 marks. As moon is simple he gives 1 mark only. Total 3 marks. Actually he had 2 younger sisters and 1 younger brother.

ELDER Co-BORNS:-

In Guru Astakavarga the Bindus contributed to the 11th house. (House of elder Co-borns) from Guru are as follows Sun 1 (being simple) Budha 2 (own house), Guru 1 (being simple)

143

Lagna 2 (being aspected by its lord) Total six. Actually he had 5 elder brothers and 1 elder sister.

ISSUES:

In Guru Binnastakavarga 5th place from Guru (Thula) has the dots of Moon, Kuja, Budha, Sukra and Lagna. The net results are: Moon-1 (simple) Kuja-6 (Retrograde and own Navamsha). Budha-2 (own house); Sukra-2 (own house). Total 13. Actually he has 13 issues. Text says about discrimination of Sex by sex of planets contributing the Bindus. But it has not been true. So I do not deal on sex discrimination.

i) In all the above tests the figures always indicate the total births and not the survivals at any period.

ii) In working out issues it is truer with Male horoscopes than of Females. Perhaps the house of children in the case of females may have to be reckoned from 9th instead of 5th. I leave it to my friends to investigate the truth.

X. SPECIALIZED BHAVA EFFECTS

The general method of delineation of merits of Bhavas such as its lord being auspicious. Bhavas and Rasis counted both from the Bhava and Lagna, the Bhava lord aspecting the Bhava and Lagna, Benefics being posited on either side of the Bhava or in Kona from the Bhava. The Bhava being aspected by functional and natural Benefics and Dishas favorable to the Bhava running in Heydays- these have been elaborately discussed in all books. To me it is only a glossary of various theories finally confusing one's mind especially when contradictory circumstances arise. I want to cut a new path and cut out only candid Truths that are applicable to all horoscopes, so whatever has proved true by research I state them only leaving the rest to my friends to judge as they strike them.

(A) GENERAL CHARACTER:-

For this Lagna is important. Aspect always includes conjunction, of Benefics on Lagna makes the native good-natured, Satwics in temperament and of best of manners. Among Benefics Jupiter's aspect is the best as that makes the native imbibe the real qualities of satwaguna. Religious mindedness, Respect and Reverence to elders, preceptors and Shastra's, which are the natural traits of Jupiter, Venus, though a benefic, gives a different effect. It tends to make one pompous, showy, sensual, Voluptuous, earthy, while Jupiter pulls up a man by purity of thought, openness of mind and Virtuous action. Here I wish to draw a line of distinction between Guru and Sukra. The former is preceptor (Acharya) of Devas while the latter of asuras. One is open-hearted and straight in his dealings while the other plays all parts (Maya). Sukra has the knowledge of making dead alive again. He has full of Indra-Jala Vidya. So will be those controlled by Sukra. Jupiter, a blunt straight forward and honest one-so his followers will be. Benefic Buddha's aspect on Lagna makes one intelligent while Full Moon though gives good effects makes one inconstant.

Malefics aspecting Lagna Make one bad character, base, mean and one of low morals and scruples. Sani is an exception in that with all the Thamasic nature in him he becomes Philosophic, Aspect of Mars makes one rushy, hot-tempered and martial.

The more the number of planets aspecting Lagna the varied are his nature and Activities in life. If no planet aspects Lagna, it is dull and vegetative life with no notable events in life. So to keep one vibrant and active it is necessary that there should be many aspects on Lagna.

(B) **FINANCE:-**

Instead of dabbling with the multiple general enunciations and combinations of planets which give no definite reading I narrate a special test to measure the degree of wealth 30, 16, 6, 8, 10, 12 and 1 are the Rays of planets from Sun onwards in order, Add the number of rays of the lords of 9th from Lagna and Moon and divide the sum by 12 and find the remainder and count so many signs from moon (Rasi). That sign becomes the Wealth Indicator. In classical language it is known as "Indu Lagna" The planet, occupying or aspecting this Indu Lagna or the lord of this sign, confers wealth during their periods commensurate with their strengths.

One text discriminates the Quantum of wealth as follows:

a) If a powerful Benefic is there or aspects Indu Lagna he earns in millions.

b) A combination of a powerful Benefic and Malefic give Lakhs.

c) If a powerful Malefic joins or aspects- Lakhs.

d) Conjunction of weak Benefic- Thousands.

e) Conjunction of weak Malefic- Hundreds.

To commonsense this does not appear to be logical. Let the Readers test the veracity of this theory. To me another theory seems to be more rational.

Sun	Karaka of	Thousands	(Sahasra)	30
Moon	Karka of	Lakhs	(Lakhsha)	16
Kuja,	Karka of	Hundreds	(Shatha)	6
Budha	Karka of	Crores	(Koti)	8
Guru	Karka of	High	(Sarvadi Raja)	10

Sukra	Karka of	Far Higher	(Shanka)	12
Sani	Karka of	Little	(ALPA)	1

These are the figures at their highest exaltation points and Zero at lowest depression point. The intermediary position have to be worked out by proportion. Additive and subtractive qualifications of the planets based on their Shadbala strength as narrated in para IX Supra should be reckoned.

Even ignoring the above two measures it is always true that if the planet in question be very powerful it confers proportionately high degree of effects during its Disha.

NOTES:

9TH house connotes one's final wealth Like-wise 2nd and 4th represents paternal wealth, professional income (Salary) from 10th and easy money from 11th.

For example in chart No. 1 of my series to find out his pay take the lords of 10th from Lagna and Moon. They are Budha in each case. The total rays is 8+8=16 i.e. 4th Rasi from Kanya (Moon Rasi) which is Dhanush. This may be called the Indu Lagna of tenth Bhava. There is Retrograde Kuja. As per above theory Kuja is Shathakaraka (Hundreds) Being Retrograde he gives 3 times hundred i.e. Rs. 300 at his highest exaltation point viz 28° of Makara. Kuja is at 26° of Dhanush-about 30° before that point his power is reduced by 30/180x300/1=Rs. 50. So his net pay is Rs. 250 only. Actually his highest pay in his service is Rs. 250.

Likewise if you want to measure the degree of effect of the 2nd Bhava consider the lords of the second house from Lagna and Moon. If of 4th, lords of the Fourth, if 11th lord of Eleventh and so on. This shows how important are the points of Lagna and moon-One is Jeeva (life-soul) while the other is Deha (body). Any effect to be of full use should be enjoyed both by the superficial earthy body for enjoyments coupled with blissful happiness felt by Antharathma. Then only it should be a real good effect.

C) ISSUES:

The chief points I deal with here are first to know if one begets issues and if so what may be total number. Never venture to predict the number of issues without getting assured of fecundity and virility. Of all the tests one given in Phaladeepika (St 14 to 16 of chapter XII) seems to be reliable. I therefore cite it for ready reference.

FIRST TEST: - FECUNDITY AND VIRILITY TESTS

a) In the case of females.   Add together the longitudes of Moon, Kuja and Guru if the result be:-

i) Even Rasi and Even Navamsha – Fecundity is assured.

ii) Mixed– there will be children only after great effort.

iii) Odd Rasi and Odd Navamsha – no Issues

b) IN THE CASE OF MALES. Add together the longitudes of SUN, SUKRA and GURU. If the result falls in:-

i) Odd Rasi and Odd Navamsha – Virility of the male to produce off-spring is strong.

ii) Mixed-there will be issues only after great effort.

iii) Even Rasi and Even Navamsha-No Virility No issues.

SECOND TEST SANTHANA RAVI: Subtract five times the longitude of Sun from five times that of Moon. This point is called Santhana Ravi "(progeny Sun)" if the Thithi represented by the result be an auspicious one in the bright half of a month (Shukla Paksha) progeny is assured without exertion. But if it is dark half of a month (Krishna Paksha) no possibility of issue. Anyway in both halves of a month it is Thithi that counts more. On New-Moon day 30, Chidra Thithi (Chowthi 4, Shashti 6, Ashtami 8, Navami 9, Dwadashi 12 and Chaturdashi 14), Vistikarnam and Sthirakarnam

(Chathushapad, Nagava, Kimsthugna and Shakuni) there will be no issues. In such cases Shanties are prescribed as narrated below:-

a) For Chidrathithi, Vistikarna or Sthiorakarna one must worship Sri Krishna by means of Purusha Sooktha.

b) For Shashti worship God Subramanian.

c) For Chaturthi worship Nagaraj (Lord of Serpents).

d) For Navami hearing Ramayana Recital.

e) For Ashtami observe 'Shravana Vratsa' (Fasting)

f) For Chaturdashi Rudrapuja (worship of Iswara)

g) For Dwadashi poor feeding.

h) For full Moon or New Moon Day worship of manes (Pithru)

To illustrate the above theory I use again chart no. I above.

FIRST-TEST MALE

Sun's Longitude is Rasis   2- 4 - 34

Venus   1-11- 7

Guru   2 – 24 -7

Total.......   <u>6-9-48</u>

The sign got by the above additions is Thula (odd sign)

The Navamaa it represents is Dhanush (odd sign)

So the virility to produce off-spring is strong

SECOND TEST:

Moon =5-9-55, so 5 times Moon is Rasi 26-19-35

Sun=2....4....34. So 5 times Sun is Rasi <u>10-22-50</u>

Balance   <u>15-26-45</u>

Subtracting 12 sings of the Zodiac Round we have 3....26....45 or 116 degrees 45 minutes. Dividing this by 12 (being the span of a Thithi) we get 9 Thithi 8 degree, 45 min. So the Thithi is Bright Half Tenth (Shukla Paksha Dashami) –a very auspicious Combination. Again as 8deg. 45 min. is second half of Shukla Dashami the Karana is 'Kharaji' (Vide my table of Karana) which is also auspicious. Thus Paksha, Thithi and Karana are all good. So he had 13 Issues.

METHOD OF DETERMINING THE NUMBER OF ISSUES

I have already described one method by means of Astakavarga. Now I will narrate a second method narrated in jathakadesha Marga (Shlokas 18...20 CHAPTER XVI)

Between guru and lord of the fifth Bhava find the stronger one. Its net rays found as follows indicates the No. of issues.

10-9-5-5-7-8-5 are the Rays at highest exaltation points of Sun to Sani in order being zero at the opposite end (Depression point). At intermediary positions find by proportions.

Then apply moderations as follows:-

If the planet be:-

i) Retrograde or exaltation Amsa.....Treble

ii) Friendly or own Amsa ......Double

   iii) Debilitation or inimical Amsa....Reduce by 1/6

   iv) Eclipsed by Sun .... zero

Exception to Sukra and Sani who lose only a moiety (half).

P.s Here 'Amsa' means 'Dwadasamsa'

All sign positions to be looked into only in his Dwadasamasa chart (D-12)

Applying this theory to Chart No. I

We have   Rays at max   Rays net

Lord of 5 Kuja is 8-25-4   5   4 + 5/6

Guru is 2-24-.7   7   6+ 5/9

In Dwadasamsa Kuja is Retrograde in Thula. So trebling his Rays we get 14 and 1/2

In Dwadasamsa Guru in Meena (own Amsa). So double it we get 13 and 1/9

Between Guru and Kuja the former is more powerful. So we have to count on Guru. So the native had 13 issues leaving off the fraction.

(D) WIFE and MARITAL MATTERS

No one Rule has so far proved to be uniformly applicable to all horoscopes in the settling of the number of marriages one may have. In Sanketha-nidhi it is said that if the lord of the Seventh Bhava and the cusp of the Seventh Bhava fall in the Navamsha of the following sets of planets (1) Budha and Sani, (2) Kuja and Sani, then he will have only one marriage. This has been no doubt found to be true but is this all the combinations? What happens under other sets of combinations? When a rule is said it must be wholesome and exhaustive what is the good if a tail —end of something is said leaving others to riddle out instances as such a combination may be rarely found. With this in view holding on the same basic principles I applied to a number of cases and have come to a conclusion which is tentative till better readings are had. And they are:-

If the two sets of planets as per the they above theory be:-

Sun-Sun	Mars-Mercury	Mercury-Jupiter
Sun-Jupiter	Mars-Jupiter	Venus-Saturn
Sun-Saturn	Sun-Venus	

You may read more than one marriage. Among these Venus-Saturn combination is a surer indication.

(p.s.) If either of these planets be Birth yogi the evil is remedied. In such cases do not predict multi-marriage.

In the above cited Chart No. 1. Lord of 7th Bhava is Budha who is in Buddha's Navamsha and the Seventh cusp of 11°-7^{1} of Mithun falls in Makaramsha of Sani. Since the combination is Budha-Sani he has only one marriage.

TO FIND THE NUMBER OF WOMEN ONE MAY JOIN

I have already furnished a method under Astakavarga (Karakath Bhava phal) to know the number of women one may join provided it is not a life of purity. I now quote another method. Note the Rays of Rasis and planets. From Mesha to Meena is order the Rays are:-

7-8-5-3-7-11-2-4-6-8-8-27.

Similarly, from Sun to Rahu in order the planetary Rays are:-

5-21-7-9-10-16-4-4-respectively.

Rule:-Find the strongest planet in the Seventh Rasi. His Rays indicate the required number.

If there is no planet the Rays of the Seventh Rasi gives the number.

P.S. These are subject to additions or reduction as per Rules stated under Karakath Bhava.

IMMORALITY:

Among many positions that cause immorality I cite a particular Rule quoted in text which proved true. In a male horoscope if Moon and Sani combined are in 7th or 9th the partner becomes immoral even at the instigation of her husband,

MANGALYAM: (WOMANHOOD OR WIDOWHOOD)

As stated before, this is rather a difficult question to answer in horoscopes. I therefore recommend the Readers to the study of marriage lines described in palmistry. I have already discussed this in detail.

E) STHOOLA BHAVA VERSUS SOOKSHMA BHAVA

Sometimes during the period a planet we experience certain effects over which it does not seem to exercise any connection by way of situation, aspect, and ownership or by natural characterstics.This will certainly puzzle an astrologer. By knowing the existence of Sookshma Bhava (minutest point of Bhava) much of the puzzle may be easily solved. Just as the effects of a Bhava are revealed by the condition of the sign occupied by the Bhava cusp the lord of the star occupied by the Bhava cusp is also functional planet of the Bhava. Thus there are two functional planets (Adhipathya) for a Bhava the Rasi lord is Sthoola (Broad based) while the starry lord is Sookshma (minutest). Both give the effect of the Bhava during their periods. In fact before the conception of Rasis (hence Bhavas) starry positions were being followed. Thus the stellar lords are as important as the Rasi lord & even more important especially when we deal with the measure of Udu Disha.

I illustrate this with my Chart No.1. (Birth 19-6-1907 at 07.09 p.m.)

Sun 4-37, Moon 9-54, Mars Retrograde 25-04, Mercury 28-26, Jupiter 24-07, Venus 11-07, Saturn 4-34, Rahu 2-17, Ketu 2-17

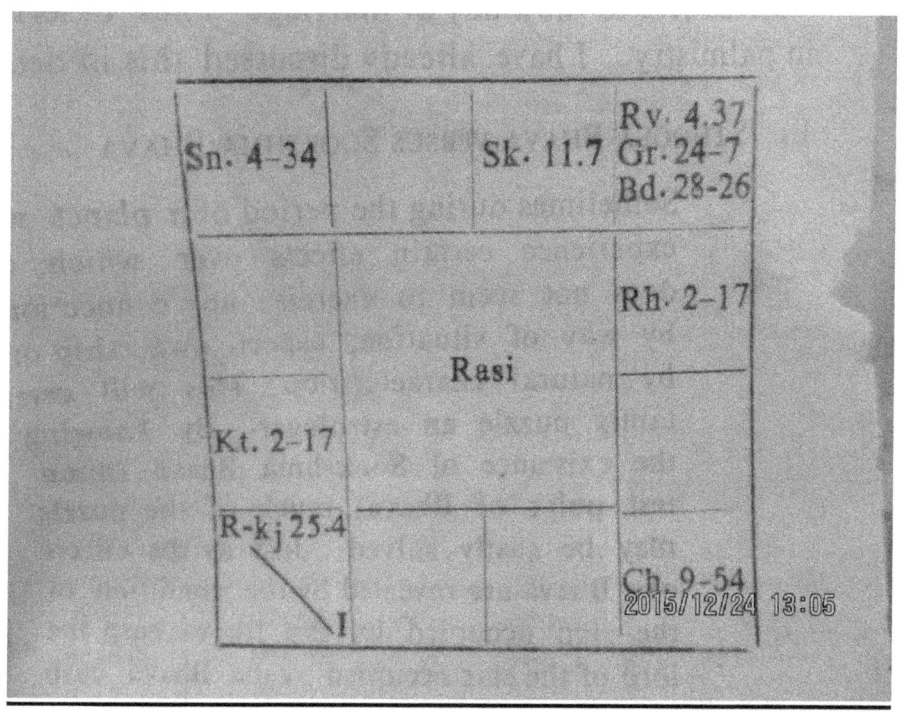

IV 17½	V 15½	VI 13¼	VII 10-57
III 15½			VIII 13¼
II 13½	Cusps		IX 15½
I 10-57	XII 13¼	XI 15½	X 17-30

2015/12/24 13:06

For Notations please see Chapter I.

From the above cuspal charts you can locate the stars in which bhava cusps are posited. The stars are

No of Bhava	Cuspal Star	Sookshma Lord	Sthool Lord
1	Moola	Ketu	Jupiter
2	Shravana	Moon	Sani
3	Shatabhisha	Rahu	Sani
4	Revati	Mercury	Jupiter
5	Bharani	Venus	Mars
6	Rohini	Moon	Venus
7	Aridra	Rahu	Mercury
8	Pushyami	Sani	Moon
9	Pubba	Venus	Sun
10	Hasta	Moon	Mercury
11	Swati	Rahu	Venus
12	Anuradha	Sani	Mars

You see from the above table Rahu and Ketu are coming to the picture though Varahmitra has ignored them. In fact they actually control some bhavas by ownership. The native was married in Rahu Disha. From Rasi Chart you cannot account for this. See above Rahu is the

Sookshma lord of 7th Bhava (house of marriage) situated in Punarvasu (Guru's star)

So his marriage in Rahu Disha Guru Bhukti or Rahu Major period and Jupiter sub period. In Sookshma Venus is the lord of 9 (father's house) situated in Rohini Moon's star (Moon lord of 8th). So his father died in Sukra Bhukti. This is not revealed in Sthoola.

The Sookshma position of the 5th Bhava is in Bharani which is female star. This Venus is again in Vrishbha a female Rasi. So he has more daughters than sons. This position cannot be gauged by Sthool position where Mars lord of 5 himself a male planet in Dhanush, a male Rasi, and all connoting male births.

Sukra gave issues during his Bhukti as lord of 5 Sookshma. You may say that children may also be read from the 11th, but that falls under alternative theory which I do not like. Moon's period as lord of 10th Sookshma has caused unhappiness to profession. You may say that as lord of 8th in 10th causes it which is true in this case. Rahu as lord of 11 has given easy money during his period or dasha. Thus you see Sookshma diagnosis gives better readings. Classifying the above we get

Sun   as   lord of   9
Moon ---------------   8, 2, 6,10
Mars   --------   5, 12
Mercury   ----------- 7, 10, 4
Jupiter -----   1, 4
Venus   --------------- 6, 11, 5, 9
Saturn ------------------2, 3, 8, 12
Rahu   ------------------ 3, 7, 11
Ketu ------------------   5, 1

MODE OF JUDGING BHAVA PHALAMS

Though I have discussed this aspect in detail at different stages yet I wish to deal with it as a recapitulatory measure before coming to a close of this Astrology Part (I) , known as Astrology made easy, for you. The following cannons may be applied always

1.   First see the planets that conjoin the Bhava or those that aspect the Bhava. This should be done from the Rasi chart reckoning the Rasi containing the cusp of that Bhava. Consider only their Natural Malefic or Benefic traits. Benefics do good while malefic spoil the effect of that Bhava.

2. Then consider both the Sthoola and Sookshma lords of the Bhava. This should be seen from the Cuspal chart and the star occupied by that Cusp. See in what Bhava these lords are posited. This should be looked into in the Bhava Chart. Judge the result as detailed before.

3. Next see in what Rasi those lords are situated. See this from the Rasi Chart. Whether he is in own house, friend's house, enemy's house, neutral house, exaltation or depression etc. Read the nature of the effects on this basis.

4. Then to estimate the quantum of effects estimate it's Shadbala. This should be seen from the Rasi Chart

5. Look to the nature of planets in conjunction with or aspecting these planets (both Functional and Natural). For good results they should not be related with their enemies. See this from Rasi chart and natural friendship etc.

6. See how many Rasis he is removed from his Bhava Rasi. Count the number of Rasis from the Rasi containing the Cusp of that Bhava to the Rasi its lord is situated. Do this in the Rasi Chart.

7. See the condition of the lord as Combust, retrograde, fast – moving or Stationary and planetary fight etc.

8. Lastly judge the modified effects of this lord from the nature of the Star in which he is placed.

    P.S   How to read the above effects are detailed in the respective Chapters.

    Considering the cannons estimate the worth of each Bhava one by one separately without clubbing together the two Bhavas owned by planet.

    With this the handling of Rasi Chart ends.

BHAVA PHALAM OF CHART NO 1

Lagna Bhava

Mars the impulsive and heaty planet is in Lagna. Saturn the Thamasic and Philosophic planet, Jupiter the Satwik and Spiritual planet, Mercury the educative and intelligent planet and Sun that throws glow in life aspecting Lagna and so these traits are seen in the life. Considering their functional traits, the effects of 5th and 12th of Mars, 2nd and 3rd of Saturn, 1st and 4th of Jupiter, 7th and 10th of Mercury and

9th of Sun will be felt as described under their Bhava effects below. As many good planets are related with Lagna, there is long life name and reputation and full of activities till the end of life.

Second Bhava

Ketu here afflicts some of the family members and cause harsh and sometimes out of the way talk out as lord of 5 does good for promotion in life and finance. Sani lord of 2 in Kendra and his own star protects finance though expenses are caused by the inimical aspects of Mars and Moon.

Third Bhava

Sani, Lord, being in 2nd from this 3rd in his own star does good to younger coborns and being in 4th from Lagna gives courage and more so by the aspect of Mars the chivalrous planet. As Rahu the other lord of this 3rd house is in 6th from this 3rd and 8th from Lagna some of the younger coborns died.

Fourth Bhava

Saturn in 4 aspected by enemies Mars and Moon disturbs personal happiness and comforts in life. Jupiter lord of 4 in 4th from this 4th and in his own star bestows the good education from one degree to the other especially good for Vedic education as Astrology being with Mercury and Sun. He promotes acquisition and pooling of assets.

Fifth Bhava

Mars lord of 5 in Lagna does good by the dent of earlier good work done. It is also good for children, but being in the star of Venus, lord of 6th is not that good to children at times.

Eighth Bhava

Rahu in 8 causes some chronic disease but being in the star of Jupiter protects often and it is Eczema. Thus his dasha is also of mixed nature.

Ninth Bhava

Sun as lord of 9th in 7th with other good planets shows acquisition of wealth through Wife – side. It is also good for general progress in life and gives high longevity to father being in 11th from this 9th.

Tenth Bhava

Moon as lord of 8 in 10 though causing sudden transfers and changes in avocation does not harm being in his own star as well the Sookshma lord of 10th. See Mercury strong with other planets that shows the bright side of profession both in service and business especially in Astrology and like fields in which there is reputation. Research is foremost seen by this combination. Saturn aspecting indicate service while Mercury as lord shows business.

Eleventh Bhava
Venus lord of 11 in 6th shows expenses while the Jupiter aspects on 11 promotes finance and that easy

ABOUT THE AUTHOR

OM GAN GANPATHAY NAMAH

Ganesh is considered as Astro Guru with Elephant head. He controls all planets or stars. I could not prove this saying till now. But I believe as Indian that if effect of planets on human beings can be proved today by me then it is also possible that I may not have gained so much knowledge to prove the above saying.

Om Laxmi Namha

Laxmi is Goddess. She is blessed by Brahma, Vishnu and Shiva. She may be known as Sarasvati (Power of Wisdom), she may be Laxmi (Power of Wealth), and she may be Kaali. Kaali is a Power of Destruction. She helps Shiva. Bhairon and Hanuman help her to perform her duty.

Om Shivaya Namah

There is no God. Brahma has power of creation. Vishnu has power of feeding all creations. Shiva has power of elimination or destruction. His third Eye is very powerful. Place of third eye may be place of energy in human head. I am not trying to advocate Hinduism. Hinduism is not a religion. It is only a way of life. It gets knowledge from Vedas. Hinduism changes very often and is not static.

There are many web sites on Astrology. They predict future on day today basis to persons born in Aries, Taurus, Gemini, Virgo, Libra, Pieces etc. It cannot be true. This is the reason mankind loses faith on Astrology. How can all the Libran suffer loss or may behave in the same way? I request my readers do not believe them as it is neither scientific nor logical. We have mastered this science. We know that to predict future one must know date of birth, time of birth and place of birth. In addition to this one must know degrees of all planets at that place of birth. Thithi is also very important. One must know division of charts. I am spending a lot of money and time on web so that these myths can be removed from the minds of many human beings of this world. I am sure people shall read my theories and shall master this subject soon.

This science needs improvement and one should study with law of probability or with statistical science as lot of Astronomy and Mathematical calculations are involved. How can one predict with numbers or with Tarot or date of birth alone? So kindly read my all web sites for this knowledge. I am writing daily on different web sites to enlighten the common person of my global world.

I am convinced and I have written many predictions and my predictions are correct. If some of my predictions are not correct then it is not due to science of Astrology but due to my mistake in calculations or my knowledge of the subject.

I have seen many web sites or books on Astrology. More I read these books, more I confuse myself. I should inform my readers that there is no need to purchase those costly books or books which contain confusing rules. I request you to read my theory or rules which are as simple as simple mathematics. For example negative multiplied by negative shall be positive or negative multiplied by positive shall be negative only. Use these simple rules while predicting future.

You cannot predict without knowing time of birth, place of birth and date of birth. Astrology is a science and is based on law of probability. It is true up to three sigma. It is only by way of research that I have been able to place before you the wonderful effects of Thithi, Weekday, Star, and Yoga etc. The Division Chart or Phala Kundali is a unique product of my long research which is foundation for correct meticulous predictions. My chief object is to spread my theory round about world.

www.JaipalSinghDatta.com   www.srimadbhagwatgeeta.com

Jaipal Singh Datta

I take pleasure in announcing to the public that I, Jaipal Singh Datta , Principal , College of Astrology, has taken a lot of pains in the research work of the noble science and has revealed very many hidden secrets, which I have practically verified as confirmed truths. I feel confident to assure readers that by faithfully and intelligently following the treatise, any one would be able to master the subject in a short time. I have written and modified writings of my Guru for the benefit of next generations.

May success crown noble efforts.